THE ROYAL MARINES

A Pictorial History 1664-1987

THE ROYAL MARINES
A Pictorial History 1664-1987

Revised Edition
(Including Gulf Conflict 1992)

Foreword by
Lt-Gen Sir Steuart R. Pringle Bt KCB DSc

PETER C. SMITH · DEREK OAKLEY

SPELLMOUNT LTD
Tunbridge Wells, Kent

In the Spellmount Military List:
The Territorial Battalions - A pictorial history
The Yeomanry Regiments - A pictorial history
Over the Rhine - The Last Days of War in Europe
History of the Cambridge University OTC
Yeoman Service
The Fighting Troops of the Austro-Hungarian Army
Intelligence Officer in the Peninsular
The Scottish Regiments - A pictorial history
Victoria's Victories
British Sieges of the Peninsular War
The 1914 Campaign

In the Nautical List:
Sea of Memories
Evolution of Engineering in the Royal Navy
Vol 1 1827-1939

In the Aviation List:
Diary of a Bomb Aimer

First published in UK in 1988 by
Spellmount Ltd
12 Dene Way, Speldhurst
Tunbridge Wells, Kent TN3 0NX

Revised Edition 1992

British Library Cataloguing in Publication Data
Smith, Peter C. (Peter Charles), 1940 -
The Royal Marines : a pictorial history
1664-1988.---(Military pictorial history
series).
1.Great Britain. Royal Marines---History
I. Title 11. Oakley, D.A. 111. Series
359.9'6'0941 VE57

ISBN 0-946771-32-4

Design and setting by Words & Images,
Speldhurst, Tunbridge Wells, Kent
Printed and bound in Great Britain by
Biddles Ltd, Guildford and King's Lynn

Frontispiece:
Mne Emmess patrols the jetty at Stromness Bay, South Georgia, as he awaits the arrival of his ship HMS Endurance. *There has been a RM detachment serving aboard the RN ice patrol ships since HMS* Protector *took over this role in September 1955. NP8901 was formed from this detachment in 1965, and HMS* Endurance *continues to carry marines.*

Foreword:
The latest Landing Craft Vehicle Personnel (LCVP Mk IV) (foreground) in company with two Landing Craft Assault (LCAs) of the Royal Netherlands Marine Corps. The LCVP belongs to the 6th Assault Squadron which is embarked in the Assault Ship (LPD) HMS Intrepid. *These LCVPs are carried on davits, whilst the LCUs are carried in the internal dock. The Mk IV is of special alloy construction and has been developed since the Falklands war. It has an enclosed steerage compartment and arcticised covers over the troop/cargo bay.*

ROYAL MARINE
ARTILLERY
1804-1923

ROYAL MARINES
LIGHT INFANTRY
1855-1923

ROYAL NAVAL SCHOOL
OF MUSIC
1903-1951

THE ROYAL MARINES

CONTENTS

FOREWORD

by Lieutenant General Sir Steuart R Pringle Bt, KCB, DSc

If it is true that a picture is worth a thousand words, this pictorial history of the Royal Marines is excellent value, especially as many of its more than 300 illustrations have not been published before. They flesh out in a fascinating manner even the most knowledgeable reader's perception of the Corps.

In case that knowledge is slender, or non-existent, the authors tell our story against their characteristically personal view of British history. Part narrative, part anecdotal, there are many new insights into the Royal Marines' past which has made us what we are. At one extreme it is interesting to read that Admiral Lord Anson played a vital role in the Corps as he is now generally seen to have done in the Royal Navy, while at the other it is illuminating, almost on the last page, to compare the Victoria Cross citations of Cpl Prettyjohn and Cpl Hunter.

Perhaps 'illuminating' is the best word to describe this excellent book, highlighting our history and, I hope, giving you as much enjoyment as it has me.

S R Pringle

INTRODUCTION

This book does not attempt to be a comprehensive history of the Corps. That would be impossible in one volume but rather it tells the story of more than 320 years of Royal Marines achievements in pictures. This has been possible through the vast fund of photographs that have been collected or donated to the Royal Marines Museum at Eastney. Just to look through more than 35,000 photographs has been a fascinating and rewarding task but to choose which ones to include has been very much a personal matter.

In making our selection we have tried to avoid those which have been over-used in other books and histories. We have also aimed to cover the whole spectrum of Corps activities, operational, social, historical and leisure. This has inevitably led to some disappointments and omissions, particularly in the colour section, where most of the Corps paintings were reproduced together in the earlier work by Peter C Smith *Per Mare Per Terram*.

The Corps are lucky to have such dedicated Royal Naval photographers serving with them who provide a constant source of excellent material, both historical and social. Many of the photographs of the past 20 years have appeared in the Corps journal *The Globe & Laurel* whom we thank and for which we make no apologies, but most of those which appeared in the magazine prior to that date no longer exist. It is to be hoped that the photographs of today will be selectively preserved for the future. The Museum archives contain a considerable number of personal photograph albums which have been donated and we have used these extensively. They are the cream on the coffee.

The text fully complements the photographs, which in themselves give a comprehensive look at changing manners, methods and styles. It updates the Royal Marines story, and takes a fresh look at some of the historical details. We have deliberately left out individual decorations except for the Victoria Cross and generally used the rank of people at the time of the incident or photograph. So many of the photographs we looked at had such sparse captions with them that it was difficult to put time and place on them let alone the names of individuals. We therefore apologise if your name was left out or spelt wrongly, but we tried hard to include as much information as possible. It is a lesson for those responsible for preserving photographs in the future.

It has not been possible to trace or acknowledge all the photographs in this book, but we would particularly like to thank Mr Harry Playford, until recently in charge of the Royal Marines Museum photographic library, Colonel Keith Wilkins (Museum Director), Miss Bridget Spiers and Mr Matthew Little of the Museum Archives and Petty Officer (Phot) Clive Dear, who copied so many of the photographs. Acknowledgement is made to the Royal Collection, the Imperial War Museum, National Maritime Museum, National Army Museum, national and provincial newspapers as well as individuals who have provided us with their personal snaps.

We are particularly indebted to Major Alastair Donald RM, who has not only vetted the text, but has contributed his considerable experience of Corps history and customs in providing and correcting much of the information used in the captions to the photographs. Special Publications of the Royal Marines Historical Society have been used extensively as well as those books included in the bibliography.

This has been a fascinating project, very time consuming but highly rewarding and we hope that our choice is not too far from your own.

Peter C Smith
Risely
Bedfordshire

MARCH 1988

Derek Oakley
Hayling Island
Hampshire

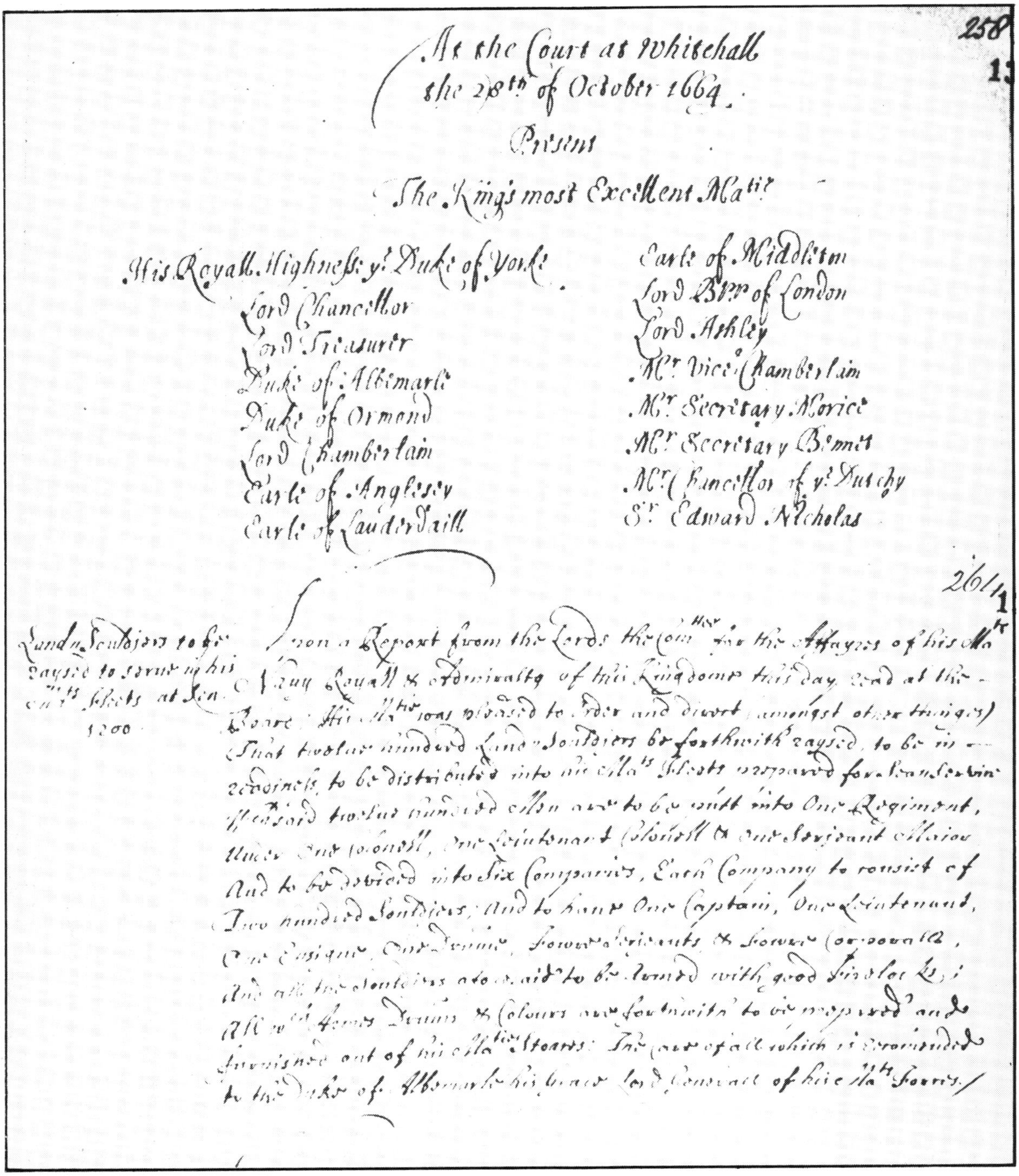

258

At the Court at Whitehall
the 28th of October 1664.

Present

The King's most Excellent Matie

His Royall Highness ye Duke of Yorke
Lord Chancellor
Lord Treasurer
Duke of Albemarle
Duke of Ormond
Lord Chamberlain
Earle of Anglesey
Earle of Lauderdaill
Earle of Middleton
Lord Bpp of London
Lord Ashley
Mr Vice Chamberlain
Mr Secretary Morice
Mr Secretary Bennet
Mr Chancellor of ye Dutchy
Sr Edward Nicholas

261

Land Souldiers to be raysed to serve in his Mats fleets at Sea. 1200

Upon a Report from the Lords the Commrs for the Affayres of his Matr Navy Royall & Admiralty of this Kingdome this day read at the Board: His Matie was pleased to order and direct (amongst other things) That twelve hundred Land Souldiers be forthwith raysed, to be in readiness to be distributed into his Mats Fleets prepared for Sea Service. The said twelve hundred Men are to be putt into One Regiment, Under One Colonell, One Lieutenant Colonell & One Serjeant Major And to be devided into Six Companies, Each Company to consist of Two hundred Souldiers; And to have One Captain, One Lieutenant, One Ensigne, One Drumme, Fowre Serjeants & Fowre Corporalls; And all the Souldiers aforesaid to be Armed with good Firelocks &c. All which Armes Drums & Colours are forthwith to be prepared and furnished out of his Mats Stoares: The care of all which is recommended to the Duke of Albemarle his Grace Lord Generall of his Mats Forces.

The Warrant of His Majesty's Court of Whitehall dated 28 October 1664 reporting that '...His Majesty was pleased to order and direct that twelve hundred Land Souldjers be forthwith raised in readiness to be distributed to His Majesty's Fleets prepared for Sea Service...'

CHAPTER 1

1664-1754
THE BIRTH AND EARLY YEARS

'The Garrison did more than could humanly be expected and the British Marines gained immortal honour.'

Triumphs of Her Majesty's Arms, 1707
(referring to capture and defence of Gibraltar)

The Royal Marines have an unmatched loyalty and dedication to the British Crown. Their period of continuous employment commenced with the establishment of the 'permanent' Divisions of the Corps in 1755, but the origins of the Marines, these famous men who are 'Soldiers and Sailors too' goes back a further 91 years to just after the Restoration of the Monarchy.

It was the outbreak of the Second Dutch War in 1664, in the fourth year of Charles II's reign, that a regiment of 1,200 men, commanded by Colonel Sir William Killigrew, was ordered to be formed in the City of London. The men were raised for service to the 'Navy Royall and Admiralty' by the Lord High Admiral himself, King Charles's brother James. They were clearly intended to be troops that could accompany the fleet although they were defined as 'Land Souldjers'. The regiment became third of the line and was known initially as the Duke of York and Albany's Maritime Regiment of Foot. The Order in Council was dated 28 October 1664. The men were strikingly arrayed in a uniform of yellow coats, red facings, sashes and stockings, said by some to be inherited from the earlier seagoing 'Firelocks' of Cromwell's day, but by others that yellow was the favourite

A Private of the Duke of York and Albany's Maritime Regiment of Foot from the painting by Charles Stadden. Although the yellow uniform may have been based on the Yellow Regiment, one of the Trained Bands of the City of London from which the regiment was formed, another more likely reason is that yellow was the favourite colour of the Duke of York. The background depicts the Tower of London.

HRH The Duke of York, the brother of King Charles II, who as Lord High Admiral, became the Colonel of the Duke of York's Maritime Regiment of Foot from 1664-1685, the forebears of the Royal Marines.

Sir Charles Lyttleton, who was Colonel of the Lord High Admiral's Regiment, the Duke of York and Albany's Maritime Regiment of Foot (later Prince George of Denmark's Maritime Regiment of Foot) from 1668 to 1689. He followed Sir William Killegrew (1664) and Sir Chichester Wrey (1665-67).

colour of the Duke of York. They were armed with matchlock muskets. The title was later abbreviated to The Admiral's Regiment.

Many of the recruits came from trained bands of the City and the formation of the regiment took but three weeks. In those days, and indeed for many years afterwards, regiments were raised for specific duties and campaigns, named after their founders and then marched off to war. Once the action was over these same men were summarily and with scant ceremony dismissed wholesale into the streets when Crown and Government had no further use for their services. Outside the campaigning season, regiments which had served afloat at such naval encounters as Lowestoft (1665) and the Four Days Battle (1666) were sent as coastal defenders to the various forts and strong-points on the east coast.

It was in such a role that the Admiral's Regiment earned one of the first of its many laurels for its stubborn defence of Landguard Fort, which had been built on the Langston spit running into the sea at the northern entrance to the River Orwell, guarding Harwich Harbour. In 1667, following the highly successful and humiliating raids by the Dutch on the Medway, four companies of the regiment were based in the town itself with two more, commanded by Capt N Darell and Capt G Cartwright, based in the fort to repel any further attempts by the Dutch fleet under Admiral de Ruyter.

The enemy duly appeared on 2 July, when eight ships-of-the-line opened a furious cannonade on the fort. They landed a 3,000-strong storming force of whom some 2,000 reached the fort, complete with twenty foot long scaling ladders, drawn cutlasses, muskets and hand 'granadoes'. The men of the Admiral's Regiment stood firm behind the fortifications and met them with firm, steady and disciplined fire. After half an hour's brisk exchanges the Dutch fell back, failing to establish a foothold on Landguard. As an eyewitness recorded:

> About an hour after they tried again, but were presently discouraged, and in a great disorder ran away, leaving some of their ladders, their hand grenadoes, and a case of very handsome pistols...

About 150 of the enemy lay dead around the fort while English losses were but three or four men, with the gallant Capt Darell wounded in the shoulder. This brief action was to prove the foundation on which an unequalled tradition was to be built in the centuries ahead.

After a short peace the Third Dutch War began in 1672, and in a pre-emptive strike the Dutch fleet met the Allied Navies in the somewhat indecisive Battle of Sole Bay. Here the regiment served afloat and lost four of its captains and four subalterns in the action. Their conduct aroused considerable admiration, a naval Capt Silas Taylor writing of 'Those marines of whom I soe oft have wrote to you behaved themselves stoutly.'

Ashore, one company under the command of John Churchill was sent to assist the Duke of Monmouth in the land invasion of Holland. In Alsace, which had been French-occupied since the Treaty of Westphalia, they fought alongside their allies at Strasbourg, but this fortress was not finally taken over by the French until 1681. Both John Churchill, later to become the Duke of Marlborough, and his two brothers served in the regiment and, after the Dutch conflicts had finished, the former went with them to the Crown colony of Virginia. This was in 1676 during a period of fierce Red Indian attacks on the settlers there.

George, Prince of Denmark. When James II came to the throne in 1685 he appointed his son-in-law the Hereditary Prince of Denmark to command the Regiment until 1896. It was then known as the Princes's Regiment and the distinctive yellow uniforms were changed to the traditional English red but retaining yellow in their facings.

The Admiral's Regiment also served as garrison troops in the Moroccan port of Tangier, which had been Portuguese for almost two hundred years, before being presented to Charles II on his marriage to Catherine of Braganza in 1662. In 1680 they fought a pitched battle with the rebellious Moors whom they soundly thrashed. Nonetheless the port proved very much a liability and four years later England pulled out altogether.

The Duke of York resigned as Lord High Admiral in 1672 and his regiment became the Duke's Regiment. In 1685, when James succeeded to the throne, command was entrusted to his son-in-law and the title changed to Prince George, Hereditary Prince of Denmark's Regiment of Foot, or more simply, the Prince's Regiment. Their distinctive yellow uniforms gave way to the traditional English red with yellow facings. In this guise the regiment was deployed to repel the expected landings of William of Orange on the south- east coast. In fact, he landed in Devon and when other regiments went over to him William banished the Prince of Denmark's men to Huntingdon. A temporary reprieve as it proved: their loyalty and close association with James proved fatal and they were disbanded in February 1689.

The continuing need for maritime troops was soon to be felt and two new regiments were raised in 1690 following French support for James. Each was made up of fifteen companies of fusiliers, wearing red uniforms with green facings. Although raised largely to replace the old regiment in its duties, many of the veterans were no longer available to serve as replacements and indeed, only one officer, Gilbert Simon, came from the old Admiral's Regiment. As was the custom of the time, they took their titles from their Colonels as Torrington's and Pembroke's 1st and 2nd Marine Regiments.

The first action of these new marines was afloat at the Battle of Beachy Head in June, when with the Dutch, now our allies, the combined fleets under the Earl of Torrington fought against the French. It was here that the phrase 'A Fleet in Being' originated, Torrington choosing tactical retreat rather than outright combat against the odds. His reasoning, to preserve the fleet to fight another day, was rejected at his subsequent court martial. Although he was acquitted, he lost command of both the fleet and the colonelcy of the regiment. Despite his caution the regiment had suffered heavy losses in the battle.

They also saw service during the siege of Cork in

A painting by Charles Stadden depicting a Private sentinel of Prince George of Denmark's Maritime Regiment of Foot in 1687. By this time the uniform had changed from the original yellow to a red coat.

The Marines landing at Gibraltar, from Cannon's 'Records of the 4th Regiment'. This Regiment (later the King's Own Royal Border Regt), commanded by Col Seymour, were made Marines in 1703, landed on the Rock on 21 July 1704 as part of the Allied force of 1,900 British Marines and 400 Dutch. The Governor of Gibraltar surrendered on 26 July and possession was taken by Admiral Sir George Rooke in the name of Queen Anne.

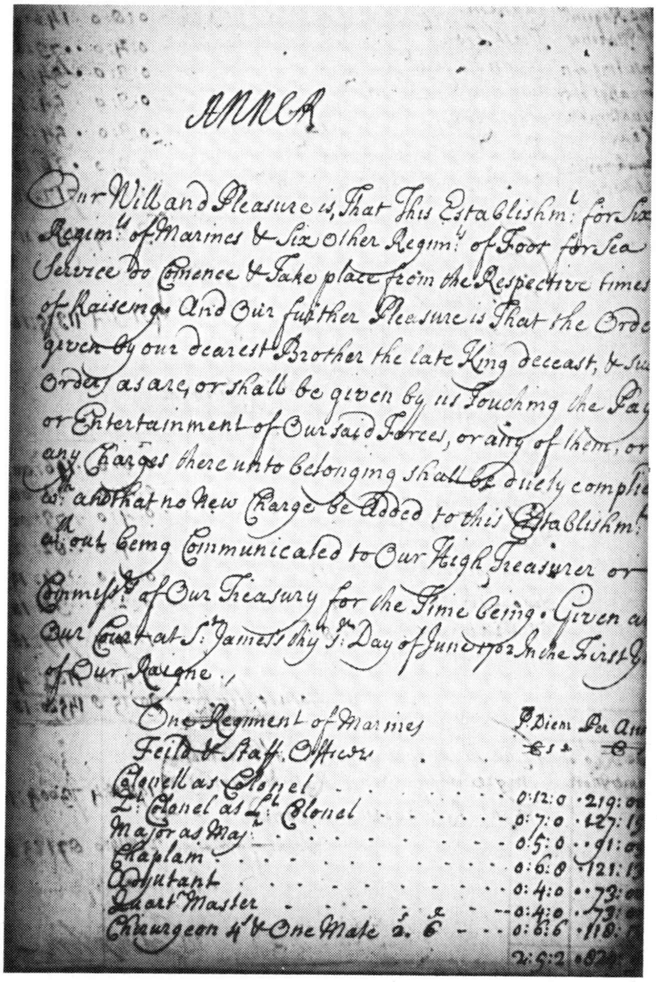

ANNE R

Our Will and Pleasure is, That This Establishm.t for Six Regim.ts of Marines & Six other Regim.ts of Foot for Sea Service do Comence & Take place from the Respective times of Raising. And Our further Pleasure is That the Orders given by our dearest Brother the late King deceast, & such Orders as are, or shall be given by us Touching the Pay or Entertainment of Our said Forces, or any of them, or any Charges thereunto belonging shall be duely complied with and that no New Charge be Added to this Establishm.t without being Communicated to Our High Treasurer or Commiss.rs of Our Treasury for the Time being. Given at Our Court at S.t James's the 1st Day of June 1702 In the First Year of Our Reigne.

One Regiment of Marines	Per Diem	Per Annum
Field & Staff Officers	£ s d	£
Colonell as Colonel	0:12:0	219:0
L.t Colonel as L.t Colonel	0:7:0	127:1
Major as Major	0:5:0	91:0
Chaplain	0:6:8	121:1
Adjutant	0:4:0	73:
Quart Master	0:4:0	73:
Chaurgeon 4s & One Mate 2s 6d	0:6:6	118:
	2:9:2	8[illegible]

A letter dated 1 June 1702 and signed by Queen Anne authorising the establishment of six Regiments of Marines. These were commanded by Cols Saunderson, Villiers, Fox, Mordaunt, Holt and Shannon. At the bottom can be seen their rates of pay, giving a Colonel's as 12/- a day and the Adjutant 4/-. It is interesting to note that over 200 years later during World War II, a Captain's rate was still only 18/- a day.

September 1690, when the Jacobites were routed. An expedition to Jamaica under Farringdon was their next major action, which again proved costly in lives. Their biggest struggles, however, were with a chary and indifferent Parliament and a corrupt pay system. This tussle was to dog them for many years to come. The war ended in 1697 and once again there were wholesale dismissals and amalgamations. Both regiments were briefly combined under the command of Colonel Thomas Brudenhall, and Seymour's, Colt's and Mordaunt's Regiments became marines, each with a strength of 754 all ranks. However, in May 1699, complete disbandment took place for the third time.

After another break, Queen Anne was again forced to re-form six 'Marine Regiments for Sea' in June 1702. This was part of the initial preparations for what was, eleven months later, to become the War of the Spanish Succession. Prince George was still Lord High Admiral but Colonel Seymour was made Super-intendent of the new Marine regiments, which contained elements of survivors of the earlier marine units. These regiments were commanded by Colonels Fox, Villiers, Shannon, Mordaunt, Holt and Saunderson respectively.

The new Marine regiments first saw service in Sir George Rooke's foolhardy attempt to repeat the Earl of Essex's earlier exploits against Cadiz. This combined operation proved a complete failure, but face was saved by an opportunist strike against Vigo, which contained a Spanish treasure fleet. The Marines assisted the army and the fleet in seizing a substantial part of the horde and so returned home in triumph instead of humiliation.

Rooke himself seems not to have learned any lessons from the Cadiz disaster, and planned a further siege of that port. However, his attentions were turned to an alternative target at the suggestion of Prince George of Hesse-Darmstadt, which was to prove an altogether happier expedition. For the

Marines it was to become their crowning glory, the capture of the Rock of Gibraltar. The first landings by 1,900 English Marines, accompanied by 400 Dutch, were made on 21 July 1704. They cut off the Rock from the mainland and called upon the Governor, Don Diego de Salinas, to surrender the fortress. He refused.

A fleet bombardment on the 23rd was followed by storming parties which established footholds in the defences. After further parleying, the Spanish Governor surrendered on the 26th and Gibraltar was taken in the name of the Austrian Emperor. But not for long. Rooke soon lowered the Austrian standard and took possession of the Rock in the name of Queen Anne. The Marines themselves then had to suffer their own siege in holding on to their newly-won prize, for when Rooke was forced to depart to refit his fleet, they were abandoned with three months supplies to withstand the attacks of the large French and Spanish armies, whose ships dominated the harbour.

The Marines' conduct during the long defence of Gibraltar was far more arduous than their original conquest, but from 9 October they held out. Sorties were made to disrupt the sapping work of the enemy, and a determined surprise attack was decisively defeated by Marines under Colonel Bor, who had replaced Colonel Fox when he was killed in action. By December sickness, disease and losses had reduced the Marines' strength to less than 1,000 men, but 200 of them managed to repulse another major enemy assault against the Round Tower on 2 February 1705, and against the Landport later that month. After eight months their endurance was rewarded and the Spanish marched away leaving the fortress in Allied hands. From that day to this it has remained a fiercely loyal British possession, a fact deeply resented by mainland Spain.

Despite their innumerable victories since, the Royal Marines still proudly bear the lone battle honour 'Gibraltar' on their colours, the name representing all subsequent actions throughout the years. They had indeed earned for themselves immortal glory, but their paymasters at the Admiralty treated them no better despite attempts at reforms under the auspices of Brigadier Seymour. In the Army today, the successors of those regiments who captured and held Gibraltar, the Queen's Regiment, the King's Own Royal Border Regiment, The Light Infantry and the Queen's Lancashire Regiment have'Gibraltar 1704-5' as one of their battle honours.

After Gibraltar the six regiments continued to give good service in the war. They served in the expedition led by the Earl of Peterborough which captured Barcelona in October 1705 and, under Colonel Wills, served in the Sardinian and Minorcan campaigns. Perhaps their most outstanding achievement was the capture in 1710 of Annapolis Royal, Nova Scotia. This base had been used by the French to harass the British possessions in the crown colonies of North America and had long been a thorn in their side. Colonel Francis Nicholson commanded the 600 Marines which were part of a strong force that sailed from England by way of Boston to evict the French from this haven. The English fleet arrived off Nova Scotia on 23 September 1710 and, two days later, the Marine regiments led the advance guard which was landed close to the fort guarding Annapolis Royal itself. They marched to within 400 yards of the defences and began throwing up siege works and digging saps. However it transpired that the mere preparations for a frontal assault on the fort were sufficient to persuade the French to surrender. The Marines were left to garrison the fort when the fleet sailed away, but they soon found themselves in turn besieged by tribes of marauding redskins, hordes of whom roamed the surrounding forests. On one occ-

A Private of Fox's Marines of 1707 from the painting by Charles Stadden, with Gibraltar as a background. This was one of six regiments formed in 1702, took part in the capture of the Rock two years later and subsequently became the Duke of Cornwall's Light Infantry.

asion this resulted in the massacre of some 70 Marines who were caught unsupported in an ambush close by.

In 1713 hostilities temporarily ceased and General Marlborough was dismissed. Three of the Marine regiments were disbanded while the other three were taken back into the Line from 1709 onwards. The only tenuous link that the Marines retained for the next thirty years were four companies of Invalids. Although there were frequent eruptions of hostility between England, France and Spain over this period, there were no more major campaigns until the War of Jenkins Ear in 1738 which started when that worthy had one of his ears cut off by Spanish excise men and who then boasted that they were going to do the same to King George II. This served as an excuse for a re-vitalised England, joined by Austria, to take up cudgels once more against their European neighbours in earnest. As had happened so often before, the King initiated an Order in Council of December 1739 to raise six Marine regiments in England and four more for the American colonies. In addition to these ten, four further such regiments were later raised at home.

The majority of these units were formed by transfer from existing Line regiments, but even so the authorities were hard pressed to raise sufficient drafts. In order to provide large enough numbers of men for Commodore Anson to equip his fleet with seagoing soldiers in readiness for his proposed campaign in South American waters, some extreme measures were resorted to, the most notorious of which was the despatch of five companies of the Invalids themselves! 500 pensioners were sent down to Portsmouth, and it is recorded that all those who were capable of walking, deserted. Their places were later taken by 240 fresh-faced youths drafted in their stead. This hotchpotch of cripples and boys then sailed away to war in September 1740.

The purpose of the Anson expedition was to attack Spanish settlements on the west coast of the South American continent while another squadron, under Admiral Vernon, attacked those on the east coast. In the end these aims were largely frustrated but instead, four years later, Anson returned home having circumnavigated the globe and he was hailed as a great explorer. However few of his Chelsea veterans lived to see England again. Among the many who did

A commissioning warrant dated 10 July 1717, appointing Edward Wolfe, Esq, a Lieutenant Colonel in 'Our Regiment of Foot'.

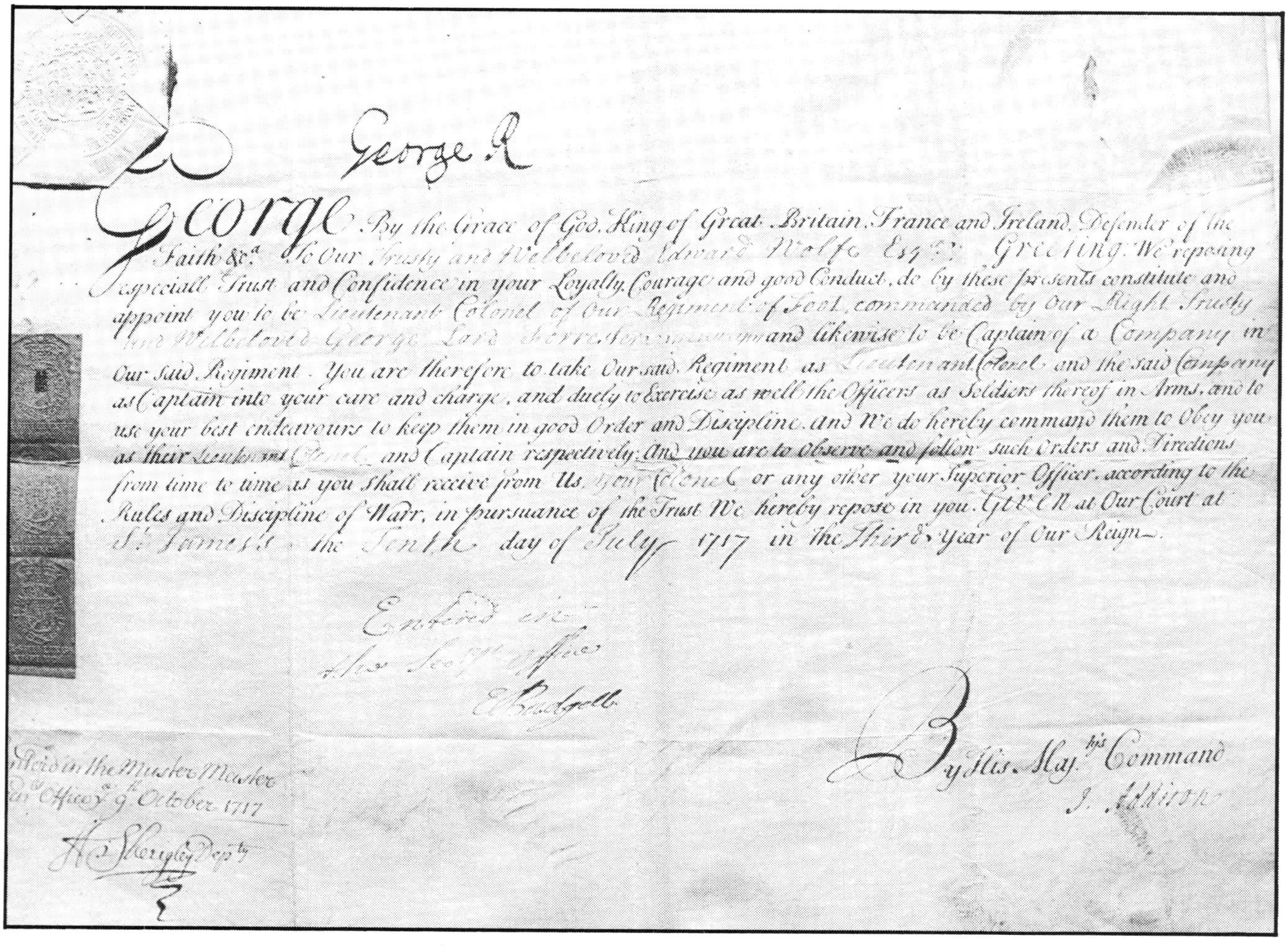

George R

George By the Grace of God King of Great Britain France and Ireland Defender of the Faith &c.a To Our Trusty and Welbeloved Edward Wolfe Esq.r Greeting. We reposing especiall Trust and Confidence in your Loyalty, Courage and good Conduct, do by these presents constitute and appoint you to be Lieutenant Colonel of Our Regiment of Foot, commanded by Our Right Trusty and Welbeloved George Lord [illegible] and likewise to be Captain of a Company in Our said Regiment. You are therefore to take Our said Regiment as Lieutenant Colonel and the said Company as Captain into your care and charge, and duely to Exercise as well the Officers as Soldiers thereof in Arms, and to use your best endeavours to keep them in good Order and Discipline. And We do hereby command them to Obey you as their Lieutenant Colonel and Captain respectively; And you are to observe and follow such Orders and Directions from time to time as you shall receive from Us, your Colonel, or any other your Superior Officer, according to the Rules and Discipline of Warr, in pursuance of the Trust We hereby repose in you. Given at Our Court at St James's the Tenth day of July 1717 in the Third year of Our Reign.

Entered [illegible] Office [illegible]

[illegible] in the Muster Master [illegible] Office 9th October 1717

By His Maj.tys Command
J. Addison

A portrait of Capt Philip Thickness c1737, who served with the Independent Company of Invalids in Jamaica in 1740. Powdered wigs were worn till about 1790.

not were Marine Cpl Crosslet and Ptes Hertford, Hobbs and Smith. They were part of the crew of the ill-fated frigate, *Wager* (Capt David Cheap) which, when the squadron was beset by furious storms off Cape Horn, became separated from the other ships on 23 April 1741. After wandering alone with her crew decimated by scurvy, the ship eventually went aground before reaching the fleet rendezvous at Juan Fernandez island.

Some 140 of the crew survived the wreck but a number mutinied and remained aboard where they embarked on an orgy of looting and destruction. When, armed with muskets, they eventually did come ashore, Capt Cheap, aided by a Marine Officer Lieutenant Hamilton, disarmed their leader and re-asserted control. As the months passed the situation proved impossible to maintain and parties left the camp to lose themselves forever in the encircling jungles; one man was shot by the Captain and order continued to deteriorate. Eventually, led by Marine Capt Pemberton, most of the crew took to the long-boats and tried to reach safety through the Straits of Magellan, leaving Hamilton and the Captain behind with a few others to make their way on foot. Disaster soon overtook the boat party for, soon after midnight on 15 December, they ran into mountainous seas that threatened to capsize the two small craft. Eventually the worst happened and the yawl was overturned and smashed. There remained but the single barge but there was insufficient room for everyone. It was therefore decided that the four Marines, named earlier, should be left behind to fend for themselves, having only arms, ammunition and a few other necessities. The Marines stoically accepted their fate but as one survivor, Midshipman John Byron, was later to relate:

> When we parted, they stood upon the beach, giving us three cheers, and called out "God Bless the King". We saw them a little after, setting out upon their forlorn hope and helping one another over a hideous tract of rocks; but considering the difficulties attending this only way of travelling left them, for the woods are impracticable, from their thickness, and the deep swamps everywhere to be met in them; considering , too, that the coast is rendered so inhospitable by the heavy seas that are constantly tumbling upon it, as not to afford even a live shell-fish, it is probable that all met with a miserable end.

As they were never seen or heard of again, we can assume this was so, but these Marines showed a brave defiance to their fate which was only rewarded with a lonely demise.

If their death was tragic then the fate of the British and American Marines that took part in Admiral Vernon's expedition to Cartegena was many times worse. Indeed it was a horrendous episode of incompetence that has often been glossed over when recalling the many heroic episodes.

The six Marine regiments with two regiments of Foot were assembled on the Isle of Wight and formed part of the 9,000 strong land force under the command of General Wentworth. Their target was the Spanish city of Cartegena in Columbia which had been chosen by higher authority and agreed by Wentworth, against the advice of Vernon, who had captured nearby Porto Bello with a small force in 1739. He warned them of the dangers of fever, disease and plague in the central American swamplands, but his first-hand knowledge was over-ruled and the whole force sailed off in January 1741 to disaster on a grand scale.

Men died all the way across the Atlantic before the expedition had even reached the tropics. 600 were put over the side before the force arrived in Jamaica, with a further 1,500 men seriously ill. Here they met up with the four American regiments commanded by Colonel Gooch, who had also suffered heavy losses through sickness and were in a state of near-mutiny, having received no orders, no food, or ammunition. Eventually, however, the survivors were re-assembled, kitted out and the whole force sailed in the middle of March.

Quarrelling between Vernon and Wentworth continued unabated and resulted in a fatal delay. If the Marines could have landed quickly, taken the town and then withdrawn, there might have been a chance of success, but this was not to be. Instead, a methodical step-by-step elimination of the lagoon's defensive forts was undertaken which duly reduced them without much effort, but at a considerable cost in time. Then instead of pressing on, it was decided that the strongest defensive position, Fort St Lazar, must be taken first. This involved the landing of more reinforcements. While he waited for these to arrive, Wentworth settled his men down among the mosquito- infested swamps. Men fell ill and some died. Soon the dreaded yellow fever began to take a grip. The reinforcements eventually joined them and, guided by locals, attacks were launched against the outlying defences. Everything that could go wrong did. The guides lost their way in the swamps. The Marines blundered around in the jungles and the enemy were fully alerted by the time the attack was eventually delivered. Grape and shot tore great gaps in the assaulting waves and only a single scaling ladder was able to be placed against the walls of the fort. Ever one who tried to climb it was killed before it was eventually thrown back. The rout was completed when the guns in the town caught the survivors in rear. The summary of this classic disaster was delivered by the commander of the 5th Marines, Colonel Grant, just before he died of his wounds: 'The General ought to hang the guides and the King ought to hang the General.'

Those who remained stumbled back to their camp and then re-embarked in their troop transports, but they were not safe even there. The fleet remained at anchor in the stifling heat of the almost land-locked lagoon for another ten days while the senior officers continued to argue about what to do next. While they wrangled, yellow fever took a firmer hold and many more Marines died. When the expedition was eventually abandoned, the fever had taken such a toll that only one-tenth of those who had sailed for Cartegena returned.

A more palatable yarn, and one that oft bears repeating, is that of the famous 'female Marine', Hannah Snell. The daughter of a Worcester hosier, born in 1723, she became an orphan before her seventeenth birthday. She had a reputation as a tomboy in her childhood and all her sisters had married soldiers. Hannah moved to Wapping and was left with child by a foreigner. She tried to trace him but after losing his trail in Coventry, enlisted in the Army. What became of her daughter is not recorded but she was said to have served in Guise's regiment and have survived 500 lashes, after which she deserted.

Her association with the Marines is believed to have begun on 27 November 1745 when she enlisted in Colonel Fraser's Regiment at Portsmouth, subsequently joining the Marine detachment aboard the sloop *Swallow*, part of the Admiral Boscawen's fleet bound for Mauritius and then on to India. Here her detachment became part of the 800-strong Marine force that joined in the attack on the French fortress town of Pondicherry, where Hannah is said to have distinguished herself during the siege and subsequent storming of the outlying forts. Stories also alleged that she was hit no less than twelve times by musket balls in the trenches, and removed the offending shot herself rather than have her sex revealed.

Her secret apparently survived intact. Whether her honour did is not recorded, but in any event she continued her service afloat aboard the frigate *Eltham*, and while that ship was in Bombay she was falsely accused of theft. The charge was dismissed and her accuser subsequently struck down by fate (or a sympathetic admirer) and so got his just desserts. Hannah herself survived to take an honourable discharge in London, later becoming, (deservedly so one must say) an actress and then the landlady of an aptly named pub in Wapping 'The Widow in Masquerade.' How much this tale owes to poetic licence is anyone's guess, but the bare facts seem to be fully authenticated, and a contemporary book on her life is in the Royal Marines Museum. Of course there have been many versions told down the years which all grew in the telling. Certainly it is a more exciting story than recent ones to have come out of Wapping.

The next expedition in which the Marine regiments took part was the capture of Louisburg on Cape Breton Island, guarding the St Lawrence river. A force of 4,000 men had blockaded the French garrison in 1744/45. The following year a naval force of four frigates from England arrived to put ashore some Marines and seamen to assist them. This reinforcement enabled the siege to be pressed ahead to a successful conclusion, with the capitulation of the French in June of that year.

In the Mediterranean they fought afloat once more in the battle off Toulon in 1744. In fact there was not much fighting and the battle became more famed for the courts-martial of the C-in-C, Admiral

Hannah Snell, the female warrior of the eighteenth century. She was born in Worcester in 1723, enrolled in General Guise's Regiment at the age of 19 and deserted. She enlisted in the Marines at Portsmouth on 27 November 1745 and was wounded at the Battle of Pondicherry, India, two years later, but despite her wounds, it was reported that her sex was not discovered. She took her discharge at Wapping on 9 June 1750 and ran a public house 'The Widow in Masquerade'

Thomas Mathews, and all his subordinates, including his second-in-command Admiral Lestock. Finding the combined French and Spanish fleets off Toulon, Mathews was duty bound to attack them, but he was so restricted by the ridiculously hidebound 'Admiralty Fighting Instructions' for carrying out each manoeuvre in battle, that he had either to allow the enemy ships to escape, or to disobey the instructions and attack only half the fleet. He courageously chose the latter course. Had fortune favoured the brave he might have become a hero, but most of his captains dared not follow his example and the result was a shambles.

On land, and much closer to home, marine regiments took part in the suppression of the Jacobite rebellion in 1745, which was intended to put Charles Edward Stuart on the throne. The Marines were present at the decisive defeat of the Scots on the field of Culloden in 1746, where they stood in the thin red line upon which the last brave and futile charge of the Highlanders crashed to oblivion.

Further sea battles took place, the first and second battles of Finisterre and the action off Belle Isle in 1747, not to be confused with the battle in 1761 for which the Corps were granted the distinction of wearing the laurel wreath. The former was much more of a decisive victory as the Marines' friend and benefactor, Admiral Lord George Anson, got round the stupid 'Fighting Instructions' simply by pretending the enemy fleet was running away and ordering 'general chase' to his own ships. The second battle in those waters was won by the dashing Rear Admiral Sir Edward Hawke with a similar ploy.

The last campaign ashore in which Marines took part during this war, was in India, already mentioned in connection with Hannah Snell. The city-state of Madras had been infiltrated by the English as early as 1644 but was captured by the French in 1746. It was re-occupied in 1748. Meanwhile a force of 5,220 men under Boscawen and Admiral Griffen, with considerable help from the East India Company, attacked Pondicherry. It had been French since 1674 and was now besieged after the 800 Marines had taken Arian Coupan, but the French held out.

With the ending of the war after the peace of Aix-la-Chapelle in October 1748, all ten English regiments were once more disbanded and their officers placed on half pay.

Admiralty Office, April 26, 1755.

HIS Majeſty having been pleaſed to appoint the following Perſons to be Officers in the Fifty new raiſed Companies of Marines; it is the Direction of the Lords Commiſſioners of the Admiralty, that they do repair, with the utmoſt Expedition, to the reſpective Head Quarters, hereafter mentioned, of the Company to which they belong, unleſs they are commanded elſewhere by their Superior Officers.

At Portſmouth, Twenty Companies.

Lieutenant Colonel Paterſon
Major Bendyſhe

Companies.	*Captains.*	*1ſt Lieutenants.*	*2d Lieutenants.*
1ſt.	Hector Boiſrond	Daniel Campbell	William Weſtcombe —— Barbar
4th.	Charles Repington	James Hill	Francis Alleſieu —— Hughes
7th.	Alexander Douglas	Donald M'Donald	Stephen Nevinſon William Goſlin
10th.	Thomas Dawes	Robert Douglaſs	Charles Templeman John Ridſdale
13th.	Thomas Moore	Colin Campbell	John Nugent Charles Champion
16th.	James Dundas	George Ord	William Sadler George Innes
19th.	John Campbell	James Short	John Purver Alexander Crawford
22d.	John Dennis	John Frazer	Tubervill Mainwright Samuel Smith
25th.	James Hamilton	Thomas Backhouſe	Robert Walſh Charles Fraſer
28th.	John Beaghan	Edward Kyffin	John Chalmers Alexander Campbell
31ſt.	Alexander Irons	William Thompſon	Peter Livingſton William Fordice
34th.	Richard Brough	James Perkins	George Preſton John Bowater
37th.	Chriſtopher Gauntlet	Pierce Dent	—— Rotheram Robert Rochhead
39th.	Walter Carruthers	Daniel Campbell	Thornhill Heathcote Benjamin Dobbs
41ſt.	William Picton	William Lutman	William Netherſole John Mackay
43d.	Richard Shuckburgh	William Rowley	Lawrence Mercer —— Alexander
45th.	George Maddiſon	Thomas Airey	Peter Campbell Hugh Lloyd
47th.	Robert Burdet	—— Walker	William Souter Peter Stuart
[illegible]	Robert Parkhurſt	—— Fell	William Lewis Adam Lodge
[illegible]	Alexander Leſlie	Enoch Markham	Charles Hughes —— Armſtrong

At Plymouth, Eighteen Companies.

Lieutenant Colonel Dury
Major Leighton

Companies.	*Captains.*	*1ſt Lieutenants.*	*2d Lieutenants.*
2d.	Gabriel Sediere	Dudley Crofts	Abraham Hilton —— Cotton
5th.	Alexander Cumming	[illegible]	Robert Johnſton
8th.	Edward Rycaut	John Suttie	—— Maltby Francis James Leſlie
11th.	John Tufton Maſon	John Phillips	Richard Mompeſſon Edward Hornby
14th.	John Gordon	Robert Ewer	Robert Mackay John Knox
17th.	George Maxwell	Lancelot Willan	Sawel Chudleigh —— Innes
20th.	Claud Hamilton	George Boſſugue	Nicholas Dunbar Robert Kennet
23d.	Thomas Dalton	Wm. Aytoun Douglaſs	—— Grant Alexander Roſs
26th.	Roger Baſket	Gerrard Dennet	Edward Farmar John Campbell
29th.	Samuel Proſſer	George Gulſton	Benjamin Leaper —— Dunn
32d.	Charles Webb	John Elliot	Maurice Wemys James St. Clair
35th.	Henry Smith	William Dennis	Samuel Barnes George Loughlan
38th.	Tooker Collins	Robert Shirley	—— Graves —— Forſter
40th.	John Vere	John Blinkhorn	Robert Chappel John Barclay
42d.	John Johnſton	Thomas Wight	—— Napier Duncan Munro
44th.	Richard Hawkins	Thomas Stamper	William Douglas John Graham
46th.	Charles Grey	Thomas Smith	Arthur Bridger Colin Graham
48th.	John Yeo	Charles Fletcher	Matthew Shaftoe —— Boſomworth

At Chatham, Twelve Companies.

Lieutenant Colonel Gordon
Major Burleigh

Companies.	*Captains.*	*1ſt Lieutenants.*	*2d Lieutenants.*
3d.	John M'Kenzie	George Langley	Styles Raven[illegible] George Norb[illegible]
6th.	Sir Rt. Abercrombie	Francis Hay	John Tupper William Dean
9th.	John Wright	Edward Howarth	Erſkine M'Kenzie Mordacai Abbot
12th.	Thomas Sheldon	John Brown	Griffith Williams —— Sullivan
15th.	Richard Barker	Archibald Campbell	Hugh Arnot Joſeph Gulſton
18th.	James Robertſon	William Frazier	John M'Fie George Brown
21ſt.	John Bell	James Mercer	Charles Mackay Robert Home
24th.	Thomas Whitwick	Denis Bond	Joſeph Smith Andrew Elliot
27th.	Henry Græme	Thomas Troy	John Shuter Archibald Campbell
30th.	Patrick M'Dowal	Richard Denniſon	Joſhua Sabine Henry Fletcher
33d.	William Stacey	William Bolton	—— Waide William Johnſton
36th.	Leathes Johnſton	Ralph Teeſdale	William Biggs William Hardinge.

By Command of their Lordſhips,

J. Clevland.

An Admiralty Office order of April 26, 1755, authorising the raising of Twenty Companies of Marines at Portsmouth, Eighteen Companies at Plymouth and Twelve Companies at Chatham, 5,000 men in all. This was the beginning of the Corps continuous existence to today.

CHAPTER 2

1755-1802
ESTABLISHMENT OF A PERMANENT CORPS

'Land Forces are nothing. Marines are the only species of troops proper for this nation. A powerful fleet and 30,000 Marines will save us from destruction'

Maj-Gen Lloyd.
The History of the Late War in Germany, 1779

The disaster of Cartagena had reflected badly on the morale of the marine regiments and they had suffered adverse public reaction to that fiasco. Admiral Anson was one who realised that the fault lay not with the brave men of those regiments but with their chaotic organisation and the poor quality of the overall administration, far worse than even the Army had known at this period. In 1747 therefore he carried through radical reforms which ensured that any future raising of marine regiments should be done solely by the Admiralty. What he envisaged was a permanent force, properly organised and based in peace and war. He had to wait until yet another conflict forced a reluctant Government into taking such a basic step; it was ever thus in the history of these islands!

Such a force must be led by regular officers, but up to this time when commissions had been purchased, the marine regiments were considered highly unfashionable for any aspiring young officer to join. A commission only rated half the cost of that in a regular Line regiment. Nevertheless when war came, the growing dominance of Britain's maritime power once more required such a force of soldiers who were capable of firing the Navy's cannons, besides being landed for shore operations.

The enemies were once more France and Spain, and the necessary Order in Council dated 5 April 1755 specified the raising of fifty companies of marines for direct service under the Admiralty just as Anson had insisted. A total of 5,000 men were to be organised into three Divisions based upon the main naval bases of the fleet, Chatham, Portsmouth and Plymouth. That they were to be 'permanent' establishments was ratified by an Act of Parliament of the same year, and thus the final step in the long road to the founding of the Marines was finally taken. The three original Divisions as allocated in 1755 became the fixed establishments of the Marines for the next two centuries, and were to be known as 'The Grand Divisions'. By the end of the war the Marines numbered almost 20,000 men in 135 companies, so great had been the need for such a versatile force in the maritime strategy adopted by Britain.

The raising of such a large force meant re-enlistment for many veterans and the drafting-in from existing Line regiments as of old. It also meant the recruiting parties scouring the land with promises of high adventure, booty and a good living. Gradually the numbers were raised and the Divisions became fleshed out with the usual mixture of the old and the new. International incidents and alarms slowly escalated into what was to be known as The Seven Years War, in which Britain again entered the lists unprepared. She recovered magnificently from early blunders to earn herself the first fruits of empire through great victories on land, and more usefully, at sea.

Among the initial disasters was the failure of Admiral Byng to relieve Majorca, whose gallant

A painting by Charles Stadden of an Officer of Marines in 1755, the year in which the Corps was permanently established. He wears a scarlet coat with white facings and breeches and carries a Fusil with a leather sling. He is depicted in front of the main gate of the Royal Citadel on Plymouth Hoe.

garrison included 110 Marines from the Mediterranean Fleet, which resulted in his execution by firing squad after a court martial. This was a shameful episode of the establishment who used, not for the last time, a service scapegoat to hide its own inadequacies. Nor did the reverses suffered by the farcical assault on Rochefort in 1757 inspire confidence. There were too many memories of Cadiz and Cartagena revived and relived, but fortunately for the Marines they were not involved. However, at this low ebb, the hour produced a saviour in the form of the dynamic Pitt the Elder, who proceeded to bring order and planning out of the chaos.

It was in 1759, the glorious year of the 'Heart of Oak', that the tide began to turn in Britain's favour, firstly with the land victory at Minden, followed by the naval success of old Boscawen off Lagos. This was crowned by the brilliant annhiliation of the French Fleet in the treacherous shoals of Quiberon Bay, where they were pursued and destroyed by Hawke. Further afield saw the recapture of Louisburg in which Marine Capt Tooker Collins led the Marine battalion. This landing took place on 8 June in face of heavy enemy fire, but once established ashore siege batteries were landed including a battery manned by Marine gunners. These scored a dramatic success against the French warships trapped inside the harbour. On 22 July Admiral Boscawen was to write how he:

> ...observed a large explosion in the harbour which proved to be the *Entrepenant* of 74 guns, blown up by our Marines battery. She fell athwart the *Celebre* and *Capricieux* of 64 guns and burnt them, and at half past eight observed a large smoak *(sic)* to arise which proved to be the barracks on fire.

Four days later the garrison surrendered and with it the whole of Cape Breton and St John islands, the latter being re-named Prince Edward Island. 8,500 prisoners were taken along with 216 guns, and eleven French warships were either burnt or sunk, all for the loss of 525 British officers and men. The path now lay open to the St Lawrence river and a year later followed the battle of Quebec and the conquest of French Canada. Marine battalions totalling 1,400 officers and men served on the peripherals of this campaign, with those landing from Admiral Saunder's fleet taking part in the action at Montmorenci Falls, and Major Boisrond's battalion serving in a diversionary role during the scaling of the Heights of Abraham by General Wolfe. Indeed Wolfe had held a commission in the Marines for a very short time.

In more distant theatres of war, 270 Marines formed part of the force which captured Manila in the Philippines in 1762, while in India Pondicherry was again taken under assault, this time successfully. Included in this force were Marines landed from the ships of Admiral Pocock's squadron. There were ample French and Spanish possessions hanging on the vine and ripe for plucking, which the British picked off one by one. Lt-Col Rycaut landed with 800 Marines to participate in the taking of the island of Guadeloupe, and they also served with Capt Swanton's force which captured Fort Royale, Martinique, on 3 February 1762. Soon afterwards they accepted the surrender of that rich French jewel.

Arthur Tooker Collins, who became Colonel Commandant at Plymouth from 1784-93. He fought at Louisburg in 1758, Belle Isle in 1761 and commanded a Marine Battalion at Boston in 1775-76. His son, Lt-Col David Collins became the first Governor of Tasmania. When he died he left a note with his will which read 'My Dear Children, considering my rank and public service for upwards of 50 years, I shall leave but a small pittance behind but be assured it is honestly got which I hope will be truly valued, there not being extortion or one single penny of dirty money in it.'

On 20 May 1765 the Marines took up quarters in Clarence Barracks, Portsmouth, which had originally been erected in 1613 as the King's Cooperage and Brewery. These barracks billeted 1,000 men and the actual site was behind the present Naval Club between Pembroke Road, St Nicholas Street and Penny Street in Old Portsmouth.

Two battalions from Admiral Pocock's squadron also helped in the taking of Havana, Cuba, from Spain in 1762.

The French suffered equally at home, when Admiral Rodney bombarded Le Havre, and although the action at Basque Roads was one of Hawke's less successful forays, it did involve the landing of a Marine battalion to reduce the French guns on the Isle d'Aix. It was in 1761 however, that the Marines achieved the second of their most abiding victories with the capture of the strategically important Belle Isle, midway between Bordeaux and Brest, off the Brittany coast.

An invasion fleet under Commodore Keppel arrived off the island on 7 April, carrying with it two battalions of Marines; one specially-raised battalion of 800 men from Chatham under Lt-Col J Mackenzie, and another consisting of 600 Marines from the fleet. In the transports were eleven Army regiments commanded by Maj-Gen Hodgson. A previous attempt the year before had been abandoned and the French were known to have reinforced the garrison since that time. The initial assaults against Port Andro to the east and another against Sauzon to the north-west of the fifteen-mile-long island, were both repulsed. Thus, in order to create a diversion while the Army tried again on 22 April, Admiral Keppel sent ashore a company of Marines at the eastern end of the island. They scaled the La Maria cliff face and, having reached the top, they and the 19th Foot stood firm against French attacks and were reinforced by a further Marine battalion and some more Army units. The gallant Mackenzie along with Capt C Murray were the only Marine officers wounded, and the former's place was taken by Capt Collins, newly promoted, and he recalled:

> I had fourteen of my men killed and wounded: and I myself was the first Marine ashore for which I am made Major, and at present command 800 men; and have the pleasure to hear everyone speak of the Marines as heroes.

Being the first ashore the Marines had

established what was to become a precedent and from that time onward the Sea-soldiers took over from the traditional Line regiments' flank companies this dangerous role, as befitted maritime warfare. Having got a firm toehold, the British reinforced their success and the bulk of the Army followed them ashore. The French were forced to withdraw to the main town of La Palais which was then besieged. In the hard fighting for the town, the Marines were heavily involved, being formed into three companies totalling 1,080 officers and men with a further 346 in detachments in the various outposts set up around the perimeter.

From time to time the desperate French garrison made strong sorties from behind their lines, and in one of these, on 3 May, Capt David Hepburn and his Marine company repulsed the enemy at bayonet point slaughtering large numbers of them. Another such attack took place on 13 May, once again with the Marines to the fore. Their fortitude carried them across several lines of trenches, clearing enemy redoubts right up to the walls of the citadel. It was recorded that 'No action of greater spirit and gallantry had been performed during the whole war.'

Eventually, on 7 June, the French raised the white flag and the town and citadel fell. The French paid the Marines the tribute of naming them *'les petites grenadiers'*, so-called because of their scarlet uniforms topped by sugar-loaf caps, while their stature did not reach normal grenadier standards. Belle Isle

Capt Wallis RN and the detachment of Marines from HMS Dolphin *are greeted by natives at Otaheite in 1767. This was one of many voyages of discovery into Polynesia and Pacific islands in which Marines took an active part.*

Capt James Cook proclaims New South Wales as a British possession at Botany Bay on 28 April 1770. He named it after its resemblance to the Glamorganshire coastline. Marines accompanied Cook on each of his great voyages of discovery.

The murder of Capt James Cook at Kowrowa in the Sandwich Islands on 14 July 1779. After one of the ship's boats had been stolen by the Hawaiian natives, Cook, with Lt Molesworth Phillips of the Marines and eight of his men, went ashore to capture Chief Teraiobu to hold him as hostage. In the ensuing fracas Cook was stabbed to death, but Phillips shot the assailant.

Engraved by A Hogg

remained in British hands until the end of the war, which terminated with the Peace of Paris in 1763. It is traditionally accepted that the laurel wreath worn as part of the Corps' badge derives from this signal victory.

The Marines' casualties amounted to 37 killed and 57 wounded. It is interesting to note that the Admiralty paymasters looked more generously upon those wounded than they had done of yore, as it is reported that a wounded officer would receive 'a gratuity of one year's full pay, reasonable expenses of his cure, and full pay while under treatment.'

On the other side of the world Marines accompanied many great voyages of exploration that took British warships to the remote areas of the Pacific and beyond. When Capt Cook was murdered in Hawaii, it was Lt Molesworth Phillips in command of the Marine escort of nine men, seven of whom were killed or wounded, who avenged his death. Despite this setback, the British colonisation of Australia was proposed in 1773, when a plan to send two companies and some artificers to pave the way for settlers was put forward. But retrenchment was once more the name of the game and the Marines were reduced to a mere 4,500 men.

Little success attended the fortunes of the Marines during the next unhappy episode in their history. Having had all the external threats to their security lifted, the colonists of America turned against the Motherland, using taxation to pay for their defence as a means of whipping up hatred and opposition. Rumours of rebellion and the storing of arms were rife and by 1775, in anticipation of trouble, further troops were sent out to reinforce the British garrisons. As so often happened before and since, British kid glove tactics, designed to overawe rather than suppress such insurrection, led not to less slaughter but to more, and ultimately to the loss of North America as a colony. The Marines played a prominent role in the opening moves of this tragedy.

Among the first reinforcements to arrive in Boston in the Spring of 1775 was a battalion of Marines. It was Major John Pitcairn who commanded the advance guard of the Provisional battalion under Lt-Col

The second advance on the Bunker Hill redoubt on 17 June 1775. The grenadier and light companies included those of the 1st and 2nd Marines on either flank. Two advances were repelled by the enemy but the third led by the 1st Marines and the 47th Regt of Foot, later the Loyal (North Lancashire) Regiment took the hill. Drawn by Howard Pyle

The fight at Concord Bridge on 19 April 1775. The Americans had established a considerable store of arms and ammunition at Concord and Light Infantry Companies from various regiments, under Lt-Col Smith of the 10th Regiment, and including Marines under Major John Pitcairn landed from boats on the Charles river to surprise the garrison.

Painting by F C Yomn

Henry Smith, despatched to Concord in April. At Lexington, Pitcairn's men were ambushed by rebel militia and the opening rounds of the American War of Independence were fired. Although some rebel guns were broken, the rebellion gained in momentum and on 20 May a further battalion of 600 Marines was landed. These two battalions, the 1st under Major James Short and 2nd under Major John Tupper, were both organised under the old veteran Pitcairn for further fighting. Each consisted of a Grenadier Company, a Light Infantry Company and eight Battalion Companies. The total ranks taking part in this campaign were 25 officers, 1 surgeon, 2 assistant surgeons, 28 sergeants, 25 corporals, 20 drummers and 600 privates.

By 17 June Boston was effectively under siege with militia guns being emplaced on the heights above the town. General Howe, the British second-in-command, scorned an amphibious landing to take these positions in the rear and opted for a frontal display of force with a set-piece advance in full order directly up Bunker Hill and straight into the enemy guns. He would have been at home on the Somme or at Ypres, one feels. His gallant officers and men certainly paid a high price for this decision which, although taken for the best of motives, was cruelly rewarded. The British force, some 2,000 all ranks, were landed below the slopes leading up to the crest in the early afternoon. Here the rebel positions were manned by fully-entrenched defenders, three times as many as the British. It was a baking hot day as the 1st Marine Battalion was paraded and dressed into line, three rows with the Marines on either flank, all the time under sniping from hidden minutemen in the houses of Charlestown. Then, led by their officers, the scarlet-coated infantry began a steady measured advance towards the palisades on top of the hill.

From their trenches the Americans poured steady fire into the massed ranks tramping steadfastly up towards them. The attackers fell thick and fast, particulary the exposed officers on whom the rebels concentrated their fire. The first advance was checked, and then the second. Re-forming their lines, still under continual fire, the advance was resumed for a third and final time. A contemporary report ran -

> a captain of Marines, who had been in several engagements before, remarked that of all the actions he had been in, this was the hottest; first from the burning of the houses in Charles Town; next the heat of the day; and thirdly from the heat of the enemy's fire. The knapsacks, therefore, were

The death of Major John Pitcairn at Bunker Hill. He is pictured (half right) cradled in the arms of his son, Lt Pitcairn (right). The Americans including Generals Putnam and Warren are depicted on the left, while other British officers include Generals Howe and Clinton (centre rear) with Major Small in the centre.
Painting by J Trumbwell

left at the foot of the slope, and some of the men even took off their coats. On the left a battalion, reduced to a handful, led the way, followed by the Marines, with the 47th in their rear.

In such steady order the disciplined troops this time reached the crest and overran the rebel lines. The enemy fired a last ragged volley and then turned and ran. One of the last shots hit the gallant Major Pitcairn who was cheering on his men with drawn sword. He fell, mortally wounded, into the arms of his son Lt William Pitcairn, who was one of the battalion's Light Infantry officers. The enemy position was broken and the rebels put to disordered flight at a cost of 450 lives. The British losses were even more punitive, over 1,045 dead and wounded lay of the field of battle. In addition to Pitcairn, the Marine dead included Major James Short, Capts S Ellis and D Campbell, Lts W Finnie, F Gardiner and R Shea and 22 Men; Capts T Averne, H Chudleigh, D Johnson and G Logan, Lts R Brisbane, J Dyer and J Ragg and 87 men were wounded. The 1st Battalion were relieved by the 2nd Battalion next day after spending the night occupying the rebel earthworks. But this bloody encounter did not break the siege and eventually the British had to evacuate the town taking all the loyalist population with them. They arrived at Nova Scotia on 4 April and Lt-Col Arthur Tooker Collins took command.

The next Marine action was by those landed from warships to help the garrison of Quebec in the repulse of the Americans led by Montgomery on 31 December 1775. Later, they sacked the rebel camp. Similarly the town of Norfolk was burnt on the first day of 1777. In April the two Marine battalions were amalgamated into one and it left for England that same autumn. Marines of the grenadier companies took part in the sacking of New York and Philadelphia, and Major Donkin in New York was to write of this time how

> This Corps, serving by sea and land, ought to have every encouragement in the power of government, but their constitution is so defective at present, that it is neither fish nor flesh! In justice to the Marines, no troops have distinguished themselves more...

Grenadier and Light Infantry headdress worn during the War of American Independence 1775. This was the first time the words of the present Corps motto Per Mare Per Terram *were seen, though its actual origination and adoption are obscure. Flora MacDonald's son, Alexander, was serving as a Marine Officer in the 6th Company of the 2nd Battalion in the battle, and it may not be coincidence that the Clan Donald motto is* Per Mare Per Terras.

They continued to do so. Marines from the fleet had participated in the unsuccessful attack on Fort Moultrie on Sullivan's Island, Charleston, on 28 June 1776 where over 200 marines and seamen were killed. At the siege of Savannah by the French in September and October 1779, the Marines from the British squadron there reinforced the garrison and were attached to the 60th Foot. Lt-Col John Maitland arrived with reinforcements on 17 September and it was he '...in command of the Grenadiers of the 60th Regiment and the Marines on the right of the line' who was responsible for repulsing an enemy attack on 9 October which resulted in the lifting of the siege.

The Marines of the fleet played a prominent role in the second attack against Charleston in 1780, taking possession of the batteries at Mount Pleasant and Lamprière's Point on 29 April, and the storming of Fort Moultrie on 5 May, leading to its capture a week later.

In 1781 Marines served with distinction in the fleet of Vice Admiral Parker, who fought the Dutch fleet under Rear Admiral Zoutman at the Dogger Bank on 5 August. They served in this capacity again at the Battle of Chesapeake when Admiral le Comte de Grasse and Rear Admiral Graves clashed on 5 September. As a result, Yorktown fell to the enemy on 19 October when many Marines were killed and wounded in the defence. It was reported that they had held a redoubt against overwhelming odds '...with uncommon gallantry.'

Fighting continued for several years, but more important considerations began to weaken British resolve. France joined in against us in 1778, Spain in 1779 and finally even Holland tried to get her share of the spoils in an anti-British alliance. Faced with threats from all the major European naval powers Britain's defences became overstretched. An increase in the financial vote to meet the needs of war saw the Marines expanded to about 20,000 men in 1781, but large Army drafts were needed to make up the required numbers of men for the fleet as well. Inevitably some of the colonies were lost, including Martinique, Guadaloupe and other West Indian prizes. Minorca in the Mediterranean went the same way and Gibraltar was only just held. It was a period of disaster in British history.

On the credit side the Dutch settlement of Negapatam was attacked. A force of 443 Marines was landed from the fleet on 21 October 1781 and the town was taken on 11 November for the loss of 13 Marines killed and 29 wounded. The following January an expedition was sent to Trincomalee in Ceylon, where a detachment of Marines under the command of Lt Samuel Orr captured the summit of a hill overlooking Fort Ostenburg. The fort was stormed by 450 marines and seamen and there were 62 casualties including Orr, who was acting as Brigade Major. Many of these distant places were thus added to the Empire, and further east the British hold over the Australasian territories also strengthened.

It was in 1786 that an earlier plan to establish a firm and lasting British presence in Australia had

The founding of Australia at Sydney Cove by Capt Arthur Phillip RN. The new Governor and his naval and marine officers about to toast the health of His Majesty King George III when the Union Flag was hoisted on 26 January 1788. The Guard of Marines were from HMS Supply.

A portrait of Lt John Hugh Griffiths who was mortally wounded in action whilst serving in HMS Prudente *at St Kitts on 25 January 1782.*

re-emerged, with Lord Sydney's authorisation of the settlement of a penal colony at Botany Bay in New South Wales. The Governor designate, Capt Arthur Phillip RN had, as part of his entourage, Major Robert Ross of the Marines as his deputy and 20 other marine officers and 192 men who had volunteered for a three-years stint, 40 of them being accompanied by their wives. They made the historic landing on 26 January 1788 and the union flag was hoisted with due ceremony and a volley fired to establish the colony. On 7 February the formal proclamation was read out at Dawe's Point by the Judge Advocate of the new colony, Marine Capt David Collins. The marines were involved with the establishment of a settlement on Norfolk Island and another at Hobart, Tasmania. Although many of the men returned home, further drafts followed in their footsteps with the arrival of the second fleet in 1789.

Meanwhile, events in France had taken a sinister turn. The revolutionary government, which considered all monarchies their natural enemies, declared war on England in 1793 and invaded Holland.

A new coalition began to emerge and the British, once more belatedly, took to their moat. Neglect had vastly reduced Marine numbers to under 5,000, and during the years needed to bring them back up to strength many Army units were sent to sea. However, by 1801, the Marines had risen to 22,696 in number and a successful Navy Vote expanded them to 30,000. They were therefore again able to play their full part in the series of devastating victories over the Republican armies, later led by Napoleon, and France's opportunist allies.

Once more the Marines were able to see combat all over the world. On 5 February 1793 Martinique was taken when Marine Capt Thomas Oldfield stormed the trenches of the St Nicholas mole and hauled down the French Colours. On 7 April the island of St Lucia fell, as did Trinidad, the two latter becoming permanent additions to the growing British Empire.

At the Battle of the 'Glorious First of June' in 1794, Lord Howe's fleet decisively routed a French fleet in the North Atlantic and marines were embarked in thirteen of his ships. The scarlet coat of a dead marine was hoisted aloft aboard the *Marlborough* when her own Colours were shot away. Between August and November the gallant Oldfield was again in action, this time in the seizure of the Dutch colony at the Cape of Good Hope, where a force of 350 Marines under Major R Hill fought at Müizenburg. On 14 August 1797 Admiral Sir John Jervis defeated a Spanish fleet, almost twice his number, at the Battle of Cape St Vincent. The Marines were heavily involved, particularly providing sharpshooters in the rigging of the ships.

The mutinies in the fleet that started at Spithead and Plymouth in April 1797 found the Marines' loyalties split. They were as often as not treated as badly as the seamen. Lord Howe's tact ensured an early end of the mutinies in June, when there was a distribution of free pardons. However, more serious trouble followed at the Nore and off Yarmouth when mutineers under the notorious misfit, Parker, soured relations although the main points of their claims had already largely been conceded. The Government treated these later outbreaks extremely seriously, and they ended with hangings and confinements to the Marshalsea. At Plymouth barracks, Marines fresh from Ireland and full of surly rebellion stirred up incidents for their own devious ends, receiving their just desserts later. Several were executed, including a sergeant hanged at the Nore. In the Mediterranean, where the mutiny spread to Sir John Jervis's fleet in 1798, the noble Lord took decisive

A Bill poster of 22 April 1792 announcing the comic opera Lionel & Clarissa *and the farce* The Prisoner at Large *performed by 'the desire of Colonel Souter and Officers of the Marine Corps' at Portsmouth.*

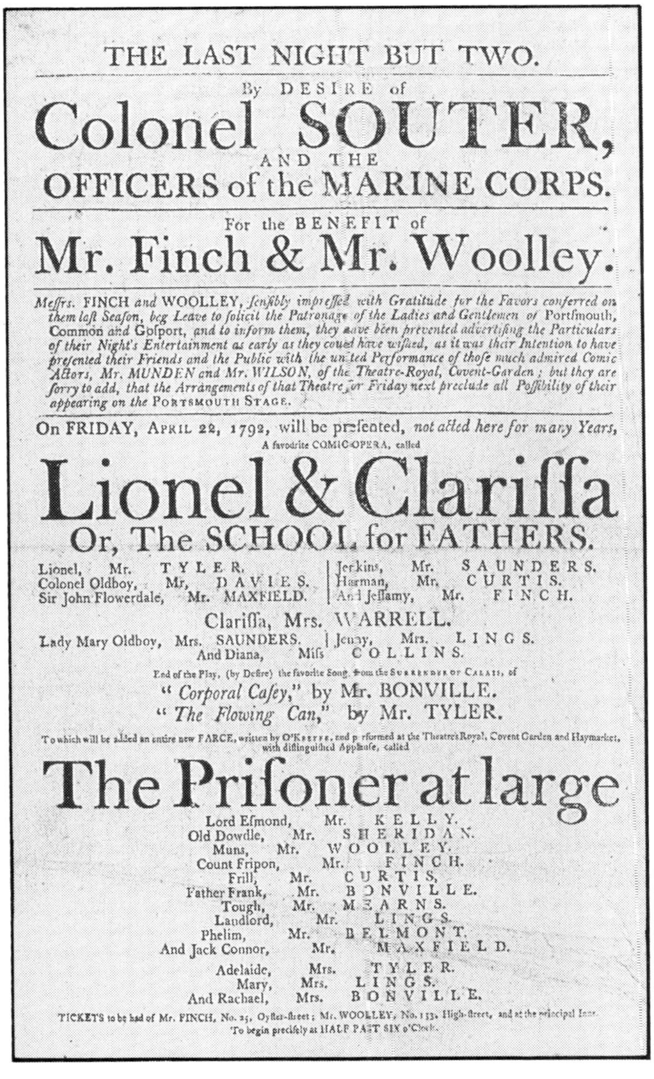
THE LAST NIGHT BUT TWO.

By DESIRE of

Colonel SOUTER,

AND THE

OFFICERS of the MARINE CORPS.

For the BENEFIT of

Mr. Finch & Mr. Woolley.

Meſſrs. FINCH *and* WOOLLEY, *ſenſibly impreſſed with Gratitude for the Favors conferred on them laſt Seaſon, beg Leave to ſolicit the Patronage of the Ladies and Gentlemen of* Portſmouth, Common *and* Goſport, *and to inform them, they have been prevented advertiſing the Particulars of their Night's Entertainment as early as they could have wiſhed, as it was their Intention to have preſented their Friends and the Public with the united Performance of thoſe much admired Comic Actors, Mr.* MUNDEN *and Mr.* WILSON, *of the Theatre-Royal, Covent-Garden; but they are ſorry to add, that the Arrangements of that Theatre for Friday next preclude all Poſſibility of their appearing on the* PORTSMOUTH STAGE.

On FRIDAY, APRIL 22, 1792, will be preſented, *not acted here for many Years,*

A favorite COMIC-OPERA, called

Lionel & Clariſſa

Or, The SCHOOL for FATHERS.

Lionel, Mr. TYLER.
Colonel Oldboy, Mr. DAVIES.
Sir John Flowerdale, Mr. MAXFIELD.
Jerkins, Mr. SAUNDERS.
Harman, Mr. CURTIS.
And Jeſſamy, Mr. FINCH.

Clariſſa, Mrs. WARRELL.

Lady Mary Oldboy, Mrs. SAUNDERS.
Jenny, Mrs. LINGS.
And Diana, Miſs COLLINS.

End of the Play, (by Deſire) the favorite Song, from the SURRENDER OF CALAIS, of
"*Corporal Caſey,*" by Mr. BONVILLE.
"*The Flowing Can,*" by Mr. TYLER.

To which will be added an entire new FARCE, written by O'KEEFFE, and performed at the Theatre's Royal, Covent Garden and Haymarket, with diſtinguiſhed Applauſe, called

The Priſoner at large

Lord Eſmond, Mr. KELLY.
Old Dowdle, Mr. SHERIDAN.
Muns, Mr. WOOLLEY.
Count Fripon, Mr. FINCH.
Frill, Mr. CURTIS.
Father Frank, Mr. BONVILLE.
Tough, Mr. MEARNS.
Landlord, Mr. LINGS.
Phelim, Mr. BELMONT.
And Jack Connor, Mr. MAXFIELD.

Adelaide, Mrs. TYLER.
Mary, Mrs. LINGS.
And Rachael, Mrs. BONVILLE.

TICKETS to be had of Mr. FINCH, No. 25, Oyſter-ſtreet; Mr. WOOLLEY, No. 133, High-ſtreet, and at the principal Inns.
To begin preciſely at HALF PAST SIX o'Clock.

steps to ensure that the Marines remained an efficient and loyal force, berthing them between the officers and the lower deck, a tradition that endures today. Both he and Admiral Duncan later commended them for their actions. This served the nation well for at the Battle of Camperdown on 11 October 1798 the Dutch fleet was defeated.

In the Mediterranean too, the Marines were well to the fore, when at Acre they landed from Capt Troubridge's squadron to serve under the inspired leadership of Admiral Sir Sidney Smith and blocked Napoleon's advance from Egypt to Turkey in 1798. On 20 March the armies of Kleber and Napoleon had joined forces to invest the town. On 7 April a force led by Major Thomas Oldfield of the *Theseus* made a gallant sally from the fortress against the enemy mine sap. The French were driven from their forward positions and the mine entrance reached. Here the Marines made a note of its direction, pulled down the supports and closed off the tunnel, carrying off the enemy trenching tools. Oldfield fell during the attack and the French paid tribute to his courage, General Bertier writing how:

> He advanced boldly towards the entrance to the mine at the head of some of his intrepid countrymen; they attacked like heroes, and were received by heroes; death only checked their bold career; the remainder retreated and took refuge in the fortress. The approaches to our parallels remained covered with the dead bodies of English and Turks. The corpse of Capt Thomas Oldfield was carried off by our grenadiers; they brought him to our headquarters; he was at the point of death, and soon after his arrival was no more; he was buried among us, and he carried with him the esteem of the French Army.

The French counter-attacked, but Major James Douglas of the Marines organised the setting-up of a British 24-pounder gun to give the defenders some artillery. Among those who also fell was Lt John Wesley Wright and Sir Sidney, whose old friend he was, asked for a volunteer to bring in his body. Marine Pte David Close volunteered and walked coolly back to the scene of the fighting. The French respecting his courage, allowed him a free passage. Close found Wright was still breathing, with two bullets in his arm with much loss of blood. He carried the wounded officer back to the British lines, Wesley Wright made a full recovery, and was later promoted to captain to become an added thorn in Napoleon's side.

In that same part of the world, the final seal was put on the little general's ambitions by the landings at Aboukir Bay in March 1801. A 600 strong Marine battalion under the command of Colonel Walter Smith executed a brilliant night march on the 12/13th to join up with the main army force under General Sir Ralph Abercromby, but suffered heavy losses in leading the subsequent attack. These same Marines also played a leading role in the investment of Aboukir Castle, and after retracing their steps they took part in the early stages of the siege of Alexandria.

At the victory over the Danish fleet at Copenhagen in 1801 by Admiral Hyde Parker's fleet, with Nelson as his second-in-command, the Marines played a lesser part. After six years of conflict the Treaty of Amiens brought about a short lull in hostilities.

During the preceding half-century the Marines had acquitted themselves so well in action throughout the world, that on 29 April 1802 King George III recognised their loyal service by directing that the Corps should be styled 'Royal', and Royal

A notice by the Commissioners for executing the Office of Lord High Admiral of Great Britain and Ireland of 12 June 1790 authorising the raising of a Company of Marines of one officer, five sergeants, five corporals, four drummers and 100 marines.

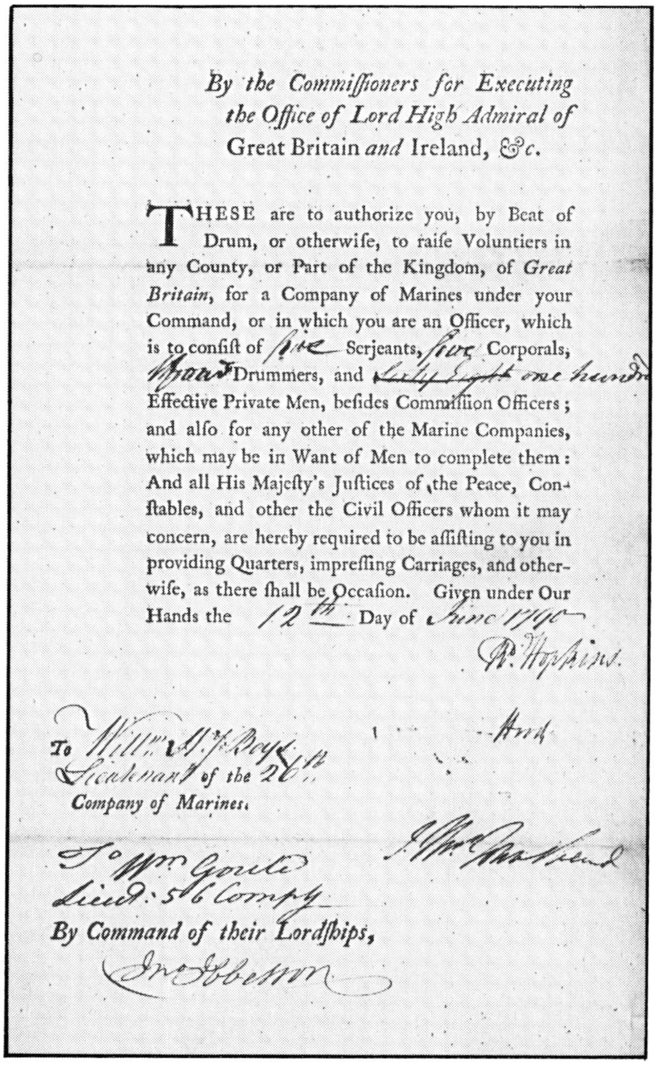

By the Commiſſioners for Executing the Office of Lord High Admiral of Great Britain *and* Ireland, &c.

THESE are to authorize you, by Beat of Drum, or otherwiſe, to raiſe Voluntiers in any County, or Part of the Kingdom, of *Great Britain*, for a Company of Marines under your Command, or in which you are an Officer, which is to conſiſt of five Serjeants, five Corporals, four Drummers, and ~~Sixty Eight~~ one hundred Effective Private Men, beſides Commiſſion Officers; and alſo for any other of the Marine Companies, which may be in Want of Men to complete them: And all His Majeſty's Juſtices of the Peace, Conſtables, and other the Civil Officers whom it may concern, are hereby required to be aſſiſting to you in providing Quarters, impreſſing Carriages, and otherwiſe, as there ſhall be Occaſion. Given under Our Hands the 12th Day of June 1790

R. Hopkins

To [illegible] *Lieutenant of the* [illegible] *Company of Marines.*

By Command of their Lordſhips,

A print of a Captain of Marines by Rowlandson first published 15 February 1799.

A letter from the Secretary to the Admiralty to Maj Gen Thomas Avarne at Portsmouth dated 29 April 1802 reporting that His Majesty (King George III) 'had been graciously pleased to direct that the Corps shall in future be styled the Royal Marines'.

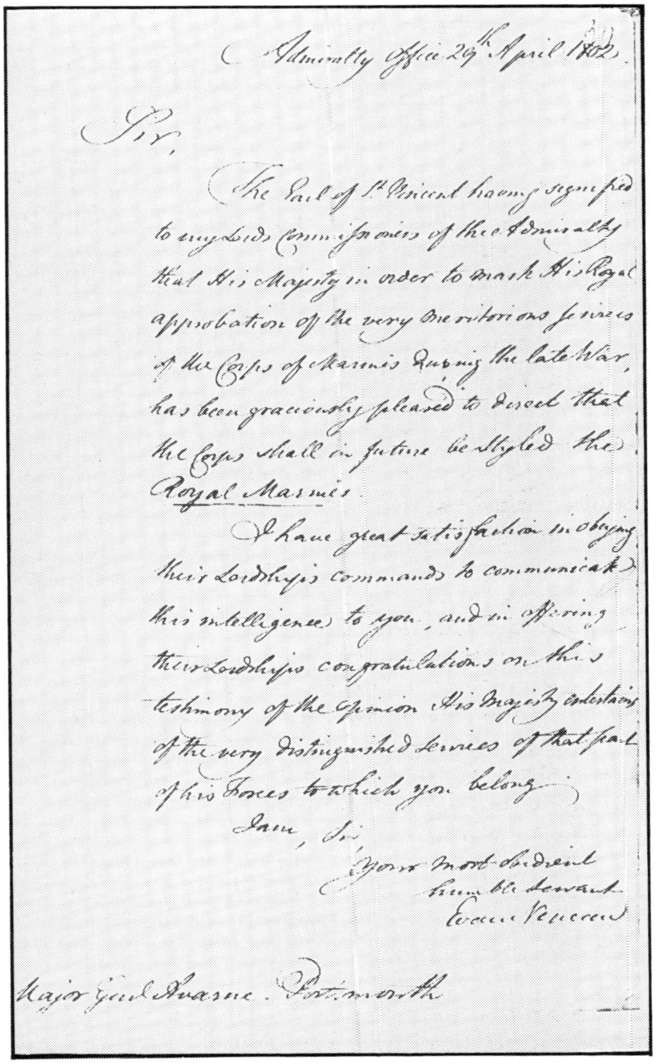

Admiralty Office 29th April 1802.

Sir,

The Earl of St Vincent having signified to my Lords Commissioners of the Admiralty that His Majesty in order to mark His Royal approbation of the very meritorious services of the Corps of Marines during the late War, has been graciously pleased to direct that the Corps shall in future be styled the Royal Marines.

I have great satisfaction in obeying their Lordships commands to communicate this intelligence to you, and in offering their Lordships congratulations on this testimony of the Opinion His Majesty entertains of the very distinguished services of that part of his Forces to which you belong.

I am, Sir,
Your most obedient
humble Servant
Evan Nepean

Major Genl Avarne. Portsmouth

Marines they have been ever since. In consequence, the white facings on their scarlet coats were replaced by the blue of a Royal regiment. Much of the credit for gaining the Corps this honour must go to Admiral Jervis, ever the Marines champion. As the Earl St Vincent and hero of the hour, he later expressed his feelings in this way:

> In obtaining for them the distinction of "Royal" I but inefficiently did my duty. I never knew an appeal made to them for honour, courage, or loyalty, that they did not more than realize my highest expectations. If ever the hour of real danger should come to England, they will be found the country's sheet anchor.

And so, to this very day, it has proved.

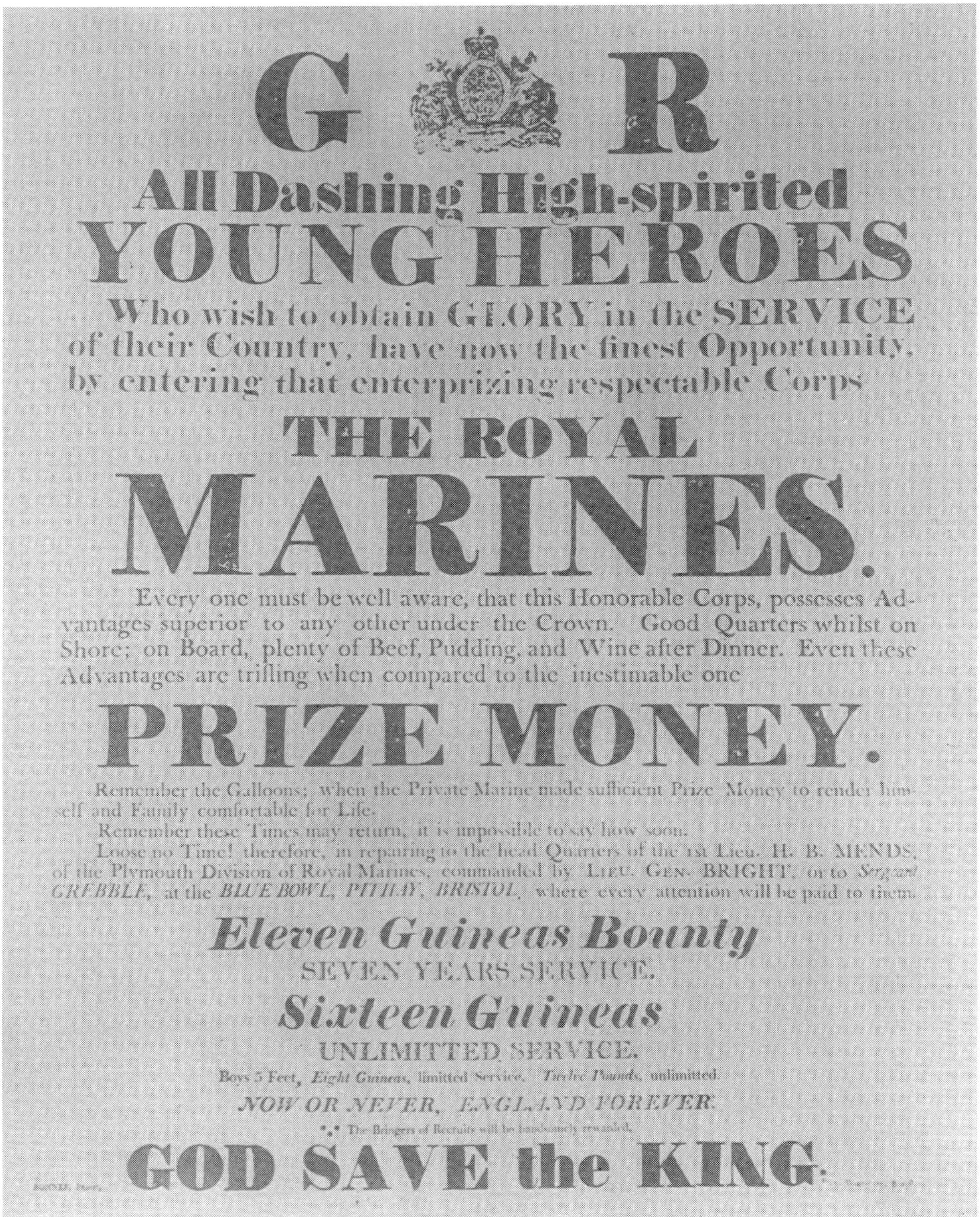

A recruiting poster for 'All Dashing High-spirited Young Heroes' c1805, soon after the Corps was made 'Royal', which offered amongst other incentives 'Plenty of Beef Pudding and Wine after Dinner' and small print saying 'The bringers of Recruits will be handsomely rewarded.'

CHAPTER 3

1803-1840

NAPOLEON AND THE YANGTSE

'His Majesty has selected for you (The Royal Marines) the badge which I this, day, by his kind permission, present to you, - a badge which you have hardly and honourably earned. From the difficulty of selecting any number of places to inscribe on these standards, your Sovereign has been pleased to give them the "great Globe itself", as their prosperest and most distinctive badge.'

HRH The Duke of Clarence, Admiral and General of Marines when presenting new Colours to Chatham Division, 26 September 1827.

The conflicts which had occurred at more and more frequent intervals, now culminated in the Napoleonic War from 1803-1815. With the conceited artillery officer-turned-General of the republic now self-elevated to Emperor, the French Grande Armèe cut swathes of destruction across the length and breadth of Europe, laying many a proud nation low, but never managing to match its land supremacy at sea. The stern bulwark of the Royal Navy thwarted Napoleon's lofty ambitions as totally as they had Phillip of Spain's in earlier years, and as they were equally to frustrate Hitler's in a later era. Command of the sea, total and absolute, not only preserved Great Britain from the ravages of such despots but enabled the deployment of British armies overseas to any point of the globe. Throughout the whole of this epic period the Royal Marines enhanced their already enviable reputation and they saw deployment on a larger scale than ever before.

Some 12,000 Marines were on the Admiralty strength at the outbreak of war and this was to increase to 31,400 by 1812. The constant use of the Royal Marines in the van of amphibious operations in the past constantly raised the vexed question of the Corps having its own artillery. Time and time again Marines had in fact manned big guns, but often as not had been forced to rely on the Royal Artillery in this respect, as well as manning the bomb ketches and other inshore bombardment craft. The question was far from new and various schemes and proposals had been made down the years to rectify this lamentable situation. For example, in 1773 Capt Wood of the Marines had proposed that practice batteries of heavy guns be provided for the Corps, and three years later the request was made again. Vice Admiral Sir Hugh Palliser, aboard the *Formidable* in Plymouth Sound, wrote to the Earl of Sandwich on 9 August 1778 about the state of the fleet and the lack of able men in all respects '...especially what concerns the gunner's matters...'. Lt-Col Arthur Tooker Collins, Palliser wrote, had his own ideas on that score

> '....he intends to offer to your Lordship's consideration a proposal for the Marines at quarters to be exercised at great guns as well as small arms. I am of the opinion it will be very useful, for every ship is obliged to take some of their Marines to man their great guns....'

Three years later, in the Colonies, Maj-Gen Patterson formed the New York Marine Artillery Company with a force of eighty men to overcome this problem locally.

The new war had not been long started when this old sore began to rub once more. It had been brought about by complaints from both the Nore and the Mediterranean that army artillery officers serving afloat were not subject to naval discipline, and that this was leading to problems. The solution was obvious and the nettle was therefore finally grasped. In 1804 the formation of Artillery Companies for the Corps took place. On 18 August an Order-in-Council proposed that the King authorise the formation of an artillery company at each of the Divisions in addition to the present establishment of Royal Marines. Thus, each of these could boast its own gunner company.

Although a fact at last, the RMA (Royal Marine Artillery) was almost stillborn, for the Admiralty was not initially enamoured of its new child and the gunners had to fight hard and prove themselves. A fourth 'Grand Division' was added with the establishment of one at Woolwich in 1805 and to complement it another RMA company came into being. They remained at this strength, despite large calls on their service during the war. The first RMA company went afloat aboard the *Devastation* in January 1805 in the squadron blockading Boulogne, while detachments of the RMA served aboard many other craft during the attack on that port in November of the same year. Here, too, they first used in anger the Congreve rocket, which quickly became firmly established with the British forces in all theatres. The Marines and Navy particularly favoured it and used it extensively in the Naval brigades in the decades that followed. The use of blue fatigue dress had long been an artillerymen's right and this became the practice of the RMA in action, but on official occasions they still wore the red infantry uniforms of their Royal Marines brothers-in-arms during the great conflict. It was not until after peace returned on 12 October 1816 that the Admiralty finally conceded 'The Royal Marine Artillery are to be clothed as the Royal Artillery except for their buttons and hats which are to be the same as those worn by the Royal Marines.' And in the following year the four RMA companies were expanded to eight.

The year of Trafalgar saw the Corps serving extensively in the ships of the fleet and they fought with great distinction when Lord Nelson crowned his glorious career with the most decisive naval battle of the century. In their traditional role as marksmen standing fully exposed on the upper decks and poops of the warships, some perched precariously higher up in the rigging, they suffered enormous casualties in the close-range pummelling and wholesale slaughter that was a naval battle of those days. Today we can but try to imagine the horrors of such carnage amid the flying splinters, falling masts and rigging, with the sizzling round and chain shot that these great vessels pumped into each other at almost eyeball-to-eyeball range. But the men performed their duty in their usual stoic and heroic manner, having been trained for it and knowing no other way. They stood and fought, were torn into bloody shreds, maimed and blinded. Those who were spared steadfastly carried on, firing carefully, repelling boarders in hand-to-hand duelling, while others helped to man the big guns and keep the broadsides rolling.

At the very outset of the *Victory*'s part in the battle, as she boldly approached the French line, a doubleheaded shot carved a bloody path through the forty Royal Marines assembled on her poop, sweeping away half of them and killing eight outright. It was about 1.30pm when the gallant Lord Nelson was struck down by the shot of a French marksman perched in the mizzen-top of the *Redoutable* at a range of about 15 yards, the musket ball passing through his left shoulder and lodging in the spine. The Admiral realised the wound would prove fatal and told Capt Hardy 'They have done for me at last'. Hardy instructed Marine Sergeant Secker, along with two seamen, to carry him below.

Despite this carnage, the men stood firm and fought valiantly, ignoring orders to take cover. 'No man went down until knocked down....' recorded 2nd Lt R Roteley later. Even so vicious fire almost cleared *Victory's* upper decks at one stage, and the enemy thought she was theirs for the taking. A boarding party was sent over by the French Capt Lucas. How-

ever Hardy and two naval officers still remained on the poop and quarter decks, along with Capt Charles Adair of the Royal Marines. They quickly summoned up their men from below and repelled the enemy. Unfortunately during the close hand-to-hand fighting Adair was hit in the neck by a marksman's musket-ball and died along with eighteen other Royal Marines. A further two Marine officers and nine Marines were wounded.

Many other Marines followed Nelson and Adair to their deaths in the other ships of the fleet in this decisive battle. The next ship astern, the 'Fighting *Temeraire*' was engaged in an equally bloody struggle in which she was dismasted, losing in the process her Royal Marines captain, a lieutenant and a score of men by the end of the day. The *Revenge*, too, took her share of the Marine losses fighting the French *Aigle*. The total Royal Marine casualties were four officers and 113 men killed, with 13 officers and 219 men wounded from a total strength of 92 officers and 2,600 Marines.

At the end of the day the French flagship *Bucentaure* bombarded by three British battleships, hauled down her flag. She was boarded by Capt James Atcherley of the *Conqueror* and five men. When the scarlet tunics of the Royal Marines appeared over the rail of the French warship, her four most senior officers presented themselves to him to surrender their command. These proved to be Admiral Villeneuve, the French Commander-in-Chief; General de Contamine, commanding the French soldiers of the fleet; Capt Prigny, the second-in-command and commander of the flagship, and Capt Majendie. Far from being overawed at such an assembly of French power, Atcherley, with true Marine phlegm, declined their proffered swords, considering it improper for one of his rank to accept such a high-powered surrender. He wished to leave that honour to his ship's Capt Pellew, but found that he had already headed off for yet another prey. Nothing daunted, Atcherley obtained the key of the French ship's magazine and secured it, placing it under the guard of two Marine sentries to prevent the enemy from blowing the ship up from under them. He then transferred the French officers to the *Mars* where they formally surrendered.

Cape Trafalgar was the last and greatest of the naval battles under sail. Napoleon no longer tried to challenge the Royal Navy's command of the sea and the *Grand Armèe*, deprived of its six hours mastery

A private of the Royal Marines in 1805 on the deck of a ship of the line. Nearly 3,000 marines served at the Battle of Trafalgar suffering heavy casualties due to their exposed positions on the upper decks.
From the painting by Charles Stadden

of the Channel, marched away in frustration to vent their spleen elsewhere. This did not mark the end of the fighting at sea by any means, although there were no more full-scale pitched battles. Spasmodic fighting was to continue for another ten years and the Royal Marines had ample opportunity to add to their illustrious record in that decade. In the aftermath of Trafalgar, for instance, a squadron under Admiral Richard Strachen defeated the Franco-Spanish squadron of Admiral Dumanoir, the 80-gun *Hero* losing her Marine officer in the fighting. But in the main single-ship and frigate skirmishes predominated, which gave the Marines plenty of action. This was coupled with frequent brushes ashore all over the globe. Here there is room only to record a few highlights of this stirring era.

In 1806 Sir Home Popham decided to strike at the Spanish possessions of Buenos Aires and Montevideo and the fleet made a landing under General Beresford on 25 June. 1,630 officers and men, mainly from the 71st of Foot with attached artillery, along with a strong naval detachment, went ashore. They took

The death of Nelson aboard HMS Victory *at Trafalgar on 21 October 1805. The Admiral is cradled by Sgt Secker, whilst sharpshooting marines can be seen on the upper deck and in the rigging.*
Painting by J Dighton - National Maritime Museum

Buenos Aires easily on 2 July, but the force was not large enough to hold the city. When the whole population rose up against the invaders, and Beresford was forced to surrender. Popham blockaded the port and also took Maldonado, on the Uruguayan coast, but had to face court-martial on his return home, instead of the honour he expected. Rear Admiral Stirling, who replaced him, stormed Montevideo, and included in his force were some 300 seamen and Marines from the fleet. Further disaster followed at Buenos Aires but unfortunately the Marines were not present.

Nearer home success followed success with Gambier's victory over the Danes at Copenhagen in September 1807. As impressive was the forcing of the Dardenelles by a squadron under Admiral Duckworth in February, a severe storm subsequently caused some losses and the surrender of the Turkish fleet was not achieved. The tongue of Napoleon's ambassador was stronger than the British Admiral's threats to destroy Constantinople. The Grand Ponte called his bluff and Duckworth had no choice but to withdraw. His squadron was duly fired upon by the forts that lined the narrows on their return journey. Heavy stone shot, weighing up to 800 lbs, smashed into the British ships from the Turkish batteries. Among the many casualties from this fiasco was Royal Marine Capt Rodolphus Kent of the *Canopus*. The Dardenelles were never to prove a happy hunting ground for British arms.

It might appear that the whole world was with Napoleon and against Britain, as the Dutch, the Spanish, the Portugese, the Danes and the Americans sided with him. But the Royal Navy was a sure shield that none of them could penetrate. Meanwhile the Marines were still active on distant shores the world over. Off Caracas, for example, the *Galatea* under Capt Sayer RN sent away a boarding party of six boats with five officers and 65 seamen and marines to try to carry the French brig *Lynx* by storm. A vicious fight ensued before the French ship was taken. Then there was a cutting-out expedition on the Spanish port of Begu by a party of 50 seamen and marines from the *Hyperion* on 7 August 1807. Lt Edward Pengelly led the storming party ashore and quickly took a battery of four French 24-pounder guns in grand style. The marines held the guns while the navy boarded three ships in the harbour and brought them safely out.

By 1808 the Royal Marines had expanded to 31,400 officers and men as the calls on their services steadily grew. Not all the incidents in which this enormously expanded force participated were glorious, it should be said, and to strike a balance some of the very few disgraceful episodes cannot be glossed over. One example was the refusal, in 1808,

I C Stadler's aquatint of a Private of the Royal Marines c1807. An Admiralty order of 28 September that year directed that men liable for foreign service were to have their hair cropped. In 1809 pantaloons and trousers were dyed a fast blue colour instead of the grey cloth shewn.

of some of the crew of the brig *Carnation*, under Commander Gregory, to fight back against the French *Palinure* in the West Indies after their commander had been killed. In the resulting court martial at Martinique, 32 seamen and marines were convicted of cowardice and sentenced to 14 years transportation to Botany Bay, while a marine sergeant was hanged at the yard-arm for the same offence. Much more typical, however, was the work done in the frigate action of the 12th April 1809 when the *Amethyst* duelled with the French *Niemen*, in which two of her Marine officers were wounded.

Equally outstanding was the role played by the Corps in Lord Cochrane's able defence of Fort Trinidad at Rosas in north-east Spain, also in 1808. Here a detachment of some 30 sailors and 30 marines from the *Excellent* and *Fame* garrisoned the citadel, and helped to defeat a strong French attack. 400 Royal Marines were landed from the covering fleet to take the defending forts on the River Mondego in August, when the British army went to support the Spanish who were now in revolt against the French. Five months later the Marines carried out a similar operation covering the withdrawal of the British army from Corunna.

In Italy the following year, however, a British landing resulted in a total rout. A convoy of transports was found hove to under the defences of the port of Palamos and a strong force of 250 marines, with 350 seamen, commanded by Capt Fane RN, was put ashore from the *Ajax*, *Cambrian* and *Kent*. They ran into a strong, fully-alerted defence and were badly mauled. Instead of retreating to the beach, where the ships could have provided covering fire, they tried to withdraw through the narrow streets of the town itself. Here they were further ambushed and cut off with heavy casualties, losing 33 officers and men killed, with 89 taken prisoner, many of them Royal Marines.

An attack on the Dutch island of Walcheren quickly took Flushing in September 1809, and it was intended to advance upon the great port of Antwerp described by Napoleon as 'a pistol aimed at the heart of England.' A battalion of Royal Marines was disembarked as a holding force at Veere. However, the naval and military commanders failed to agree, and while they dithered, disease spread. The force was decimated by malaria, and the whole operation abandoned.

It was against the Danes that the Corps won their greatest land victory of the era. In the Kattegat, the tiny island of Anholt was already garrisoned by 350 Marines under the command of Capt James Maurice RN with a detachment of 31 Royal Marine Artillerymen under Capt Robert Steele RMA. They were entrenched at Fort Yorke, a converted lighthouse used as a strongpoint and citadel. On 27 March 1811, a Danish force of more than 1,600 regular soldiers under Colonel Melstedt, supported by twelve gunboats offshore, launched an attack. The Danes succeeded in getting ashore undetected in dense fog, but despite the surprise they achieved and their superiority in numbers, they were totally routed by the Marines. The Danes attacked in two columns across the desolate sandhills but were met with a steady fire from the Marine gunners which badly sliced the first column. The second met a similar reception and both were halted. A second determined attack was then mounted by the whole Danish force, but as Steele described:

> Our guns and musketry mowed them down. Melstedt, the Commander-in- Chief fell by a musket ball and the next in command, poor Reytz had both his legs shot away by a cannon ball; and another cannon shot at the same time killed the brave Holstein, who commanded the southern column of attack.

A Danish-speaking RMA sergeant took the unconditional surrender of the northern column, upon which half the southern column also laid down their arms, while the rest ran away. Thus, for the loss of two men killed and 30 wounded, the small Royal Marines garrison took the surrender of 680 Danes, killed a further 200 and still held the island.

In the Far East, attacks were renewed on Dutch settlements in Java which had been handed back after the victories of earlier years. In 1811, for example, a fleet was sent out under Rear Admiral Stopford to land an army of 12,000 men under General Achmuty to take the entrenched camp of Cornelius near Batavia. A strong force of seamen and marines was landed to assist in this operation which resulted in a resounding victory where over 1,000 of the enemy were slaughtered and 5,000 taken prisoner.

Nearer home in the Mediterranean, a storming party was landed on Languelia, at the head of the Adriatic, by the battleships *America* and *Leviathan*. The Royal Marines stormed ashore and carried the defending battery in grand style while sixteen smaller craft were cut off from the harbour for the loss of but 16 men killed and 20 wounded.

A fresh battalion of 500 officers and men under Major Richard Williams was raised at home in 1813 and in November they, with their attached artillery, were despatched to the Tagus in Portugal to help

A recruiting poster of the Trafalgar age urging men of at least 5'6" and under 25 years of age to join. Most recruiting of this time was done in inns and public houses.

Lt Harry Hunt of the Marine detachment of the 38-gun frigate Undaunted *landed with 200 marines to storm the citadel of Cassis, between Marseilles and Toulon on 18 August 1813. At one stage he was cut off when the marines following him fell from the scaling ladder which broke under their weight, but he fended off the enemy until help arrived. This landing was typical of the small forays which marines carried out during this period.*

defend the flanks of the Torres Vedras, from which Lord Wellington maintained a precarious European toehold. They garrisoned Fort St Julien, Fort Almada and later Lisbon before being brought back home the following year. They returned in June attacking the fort at Lequitio on 20th and Fort St Ano, near Santander, on 3l July.

When America entered the lists against us in 1812 few troops or ships could be spared, which allowed them time for a few local victories. Eventually, however, some troops were mustered together and sent across the Atlantic. This included three Royal Marine battalions with attached RMA batteries, and they soon saw action. Under the command of General Beckwith, a force of 2,000 men including the Royal Marines 1st & 2nd Battalions, landed near Hampton, Virginia, in June 1813 and quickly captured it. Lt-Col Sir Charles Napier, commanding the 102nd of Foot, paid a glowing tribute to the RMA after this action saying that

> Never in my life have I met soldiers like that Marine Artillery. We suffered much fatigue and hardship, but never was seen anything not admirable in these glorious soldiers.

There was also much fighting on the Canadian border when the Americans were evicted from many of the towns they had occupied earlier. In an attack on Fort Oswego on Lake Ontario on 6 May 1814, the 2nd Battalion of 400 men, commanded by Lt-Col James Malcolm, was part of General Drummond's force which took this strong American position, held by at least 500 American regulars, at bayonet point in ten minutes. Lt John Hewett scaled the flagpole under sniper fire, hauled down the 'Old Glory' and hoisted the Union Flag. A wounded American whom the British had mercifully spared then tried to kill him, but was put to the bayonet by a Marine colour sergeant before he could do so. An attempt to cut out enemy boats at Sandy Creek with 180 marines and seamen under Capt Popham RN and Capt Silbury RN went badly awry when American militia and hostile Indians ambushed them, killing 18, wounding 50 and taking the rest prisoner.

In August 1814, General Ross and Admiral Cockburn mounted a punitive attack on Washington by way of an advance up the Patuxent and Potomac rivers. At the battle of Bladensburg on 24 August General Ross, with 4,000 of Wellington's veterans, attacked and put to flight an American army of 7,000 soldiers, 600 seamen and 24 guns under General Winter, who was entrenched there. The American President Madison was forced to witness the ignoble retreat of his army, who left behind them ten guns and fled to the shelter of Washington. The British losses were 65 dead and 191 wounded. The Marines in the Naval Brigade under Admiral Cockburn played a considerable part in this little-known victory. Washington was duly burnt in retaliation for the American sacking of Toronto, and the Potomac Bridge was destroyed before the victorious British re-embarked. Meanwhile, Capt Napier had forced the Potomac, shelled the defenders of Fort Washington and taken the surrender of Alexandria.

A new Royal Marine Battalion, the 3rd, was formed from the remaining companies of the 2nd and the Colonial Marines, whilst Colonel Richard Williams reformed the 1st Battalion. 130 Marines and seamen under Admiral Parker did not fare so well when they were landed at Chesapeake Bay. They were ambushed, losing Parker, a midshipman and twelve men killed. General Ross was also killed in the subsequent attack on Baltimore, but General Brooks took the American lines at bayonet point before the whole operation was abandoned on 13 September.

Back in Europe Napoleon's empire was collapsing. Gradually his allies, except the Americans, had

Napoleon surrendering himself 'to the generosity of the British nation' on board the battleship Bellerophon *on 15 July 1815. Having inspected the Captain's guard of the Royal Marines under Capt George Marshall, Napoleon commented on their skill and bravery in action saying 'Much might be done with a hundred thousand men such as these'.*
From the engraving by T Sutherland

turned against him one by one. He was beaten, exiled to Elba and then returned to France for one last gamble. When the final débâcle of the 'Hundred Days' ended in defeat at Waterloo in 1815, Napoleon was taken into exile once more, this time to the more remote island of St Helena. Aboard the battleship *Bellerophon*, in which he was to take passage, he had the opportunity to examine at close quarters the men who had contributed so much to his downfall. He inspected the Captain's guard of the Royal Marines detachment under Capt George Marshall, and even asked them to go through some simple drill movements. He expressed his admiration at their skill, saying that he though the French method of charging, not just with the front rank only as the Marines did, was superior, Marshall, a fluent French speaker, dryly commented that it had not proved very successful in practice, to which the little dictator could only reply

> Much might be done by a hundred thousand men such as these.

With the final vanquishing of the scourge of Europe, a long period of relative calm and tranquillity followed. Through her long and diligent efforts, Great Britain had raised herself to a position of unrivalled prestige and power. With an enormous technological revolution about to be launched by her designers and engineers, coupled with the secure and inviolate shield of the Royal Navy, Britain prospered as never before or since. The Empire expanded and the small wars of annexation, pacification and rebellion provided the Royal Marines with the greater part of their fighting during the next 100 years. Combat at sea became a rare experience.

Two of the Royal Navy's duties policing the seas during this period of *Pax Britannica* were the curbing of slavery and piracy. In both spheres she was outstandingly successful, although little credit is given these days. The first punitive expedition that involved the Royal Marines was the crushing of the Barbary Corsairs at Algiers on 27 August 1816. Representation to the Dey had been rebuffed. It was known that 1,200 Christian slaves were held in the city, which was very strongly defended with over 1,000 cannons of various sizes, eight mortars, four frigates, five corvettes and mortar boats, manned by 40,000 men. Lord Exmouth, formerly Sir Edward Pellew, commanded the British fleet sent out from Plymouth to the Mediterranean to bring the Dey to his senses. They were joined by a Dutch squadron under Admiral Van de Cappellen. Marines served in the six British battleships which carried out the initial bombardment. Because of the shallow waters, smaller craft were used for close-in work and the RMA manned these little 'bomb vessels', with RMA Lts R Henry, J Maule, F James and J P Bissett captaining the *Beelzebub*, *Fury*, *Hecla* and *Infernal* respectively. They took up position some 2,000 yards from the enemy's main batteries, and bombarded them to very good effect.

Lord Exmouth's flagship, the 100-gun *Queen Charlotte*, sent away her barge with a Royal Marines detachment commanded by Lt S A Rhodes Wolrige RMA. For the loss of two dead they clambered aboard and burnt the vessel, before returning safely to the flagship. Enemy fire from the huge number of guns caused widespread casualties and damage to the fleet, but they were finally silenced. The Royal Marines casualties totalled 26 dead and 115 wounded.

A similar action took place at the Greek port of Navarino Bay during the Greek war of Independence in 1827. The Turkish-Egyptian fleet of some 81 vessels of all types under Ibrahim Pasha, who also commanded the Sultan's land forces, was anchored. Ibrahim had landed troops in defiance of an agreement he had made with Admiral Sir William Codrington, the naval Commander-in-Chief Mediterranean. Codrington therefore followed him with five battleships and many smaller ships and, on 20 October, was joined by squadrons from the French and Russian navies. The combined Allied fleet entered the six-mile wide Navarino Bay through its narrow entrance, guarded by forts in Turkish hands, to find the Turkish fleet drawn up in a huge crescent ready for battle. No shots were fired as the Allied fleet took up station facing their opponents. When Ibraham Pasha requested them to leave, not having been invited to enter the Bay, Codrington replied that he 'had not been invited to receive orders, but to give them!' Shortly afterwards the Turks opened fire on a ship's boat from the *Dartmouth*, killing one of her officers and several of her crew. The action soon escalated, thereafter, resulting in the complete destruction of the Sultan's navy, while the British lost 75 killed, including the officer commanding the Royal Marines aboard the battleship *Asia*. As at Trafalgar, the marines had assembled on the poop of the *Genoa* and took heavy casualties from the upward firing guns of the Turkish vessels, whose shot pierced the lower decks below the waterline and erupted on the upper deck. Ten Royal Marines were killed in a few minutes before the *Genoa* had even fired a shot, and their commanding officer, Capt Thomas Moore, was mortally wounded. Lt J Miller and the few survivors were ordered to lie down before they, too shared their fate.

A portrait of Lt-Col Anthony Stransham 1825. He joined Chatham Division in 1781 at the age of 13 3/4 years. He was captured by the French in 1793 while serving in the frigate Thames. *Exchanged the following year, he fought in the* Monarch *at the Battle of Copenhagen in 1801. He retired by sale of his commission for the regulation sum of £4,500 in 1826. He was the father of General Sir Anthony Blaxland Stransham, from whom generations of Wilberforces were descended.*

Commodore Bathurst, who had received what was to be his fatal wound, lay dying below when Capt Moore was brought down. Recognising his voice Bathurst asked 'Ah, Moore, is that you?' to which the marine replied 'Yes I am wounded'. 'Ah' replied Bathurst '*fortune de guerre*'. Moore subsequently died

A group of marines at firing practice on the short range etched by William Heath c1833. Uniforms had changed little since the days of Trafalgar.

Portrait of King William IV. In 1827, as HRH The Duke of Clarence, Lord High Admiral, the future King presented Colours to all four Divisions. It was arranged that three days pay be collected from all officers to pay for these portraits which were designed by Sir Thomas Lawrence, but almost certainly painted by A Morton. On succeeding to the throne in 1830, by which time only two days pay had been collected, the King presented the Divisions with the portraits. The one pictured depicts Colours as a background whereas the RMA picture, now in the Museum, shows a mortar. The Plymouth portrait was destroyed in the Blitz of 1940.

A portrait of a young Marine officer c1830

A Royal Marines other ranks Regency Shako Plate c1830.

and was buried near Mount Etna in Sicily. A Royal Marine named Hill, stationed on the poop had both arms shot off. The poor fellow turned to Commander Dickson and said 'I hope you will allow, sir, that I have done my duty.' He was only 21 and was later heard singing and when asked why, he replied 'I'm trying what I can do at ballad singing, now that I have lost my arms.' Sadly, he too died of his terrible wounds in hospital later. In another ship, a Marine who had his arm shot off during the fight, picked it up and laid it on a shelf with the defiant words 'There's an example for you all.'

This is but one small example of the many incidents in which Royal Marines were involved during this period. They saw action during the Spanish 'Carlist' or Civil War in 1837, sending a battalion and a field battery to support the volunteer 'British Legion' which was backing the government in the fighting. They remained for three years under Lt-Col J Owen, and were instrumental in putting backbone into the Legion during the fighting at the siege of San Sebastian. Their finest hour was at the battle of Hernani on 15 March 1837, when they withstood attacks by four times their strength.

The battle commenced with a strong foray against the besieging Carlist lines by the defenders, the well-aimed delivery of the spherical case-shot from the RMA battery making the biggest contribution. The first line was taken with comparative ease, and the River Urimea crossed. Their infantry brothers-in-arms were called upon to show their steadfastness the following day, when the Legion was broken into panic by a Carlist counter-attack. Only the firmness with which the Marines held their line prevented a total rout and defeat. An eyewitness recorded: 'What a fine example of discipline the Marines gave! Had they not acted as they did, our right would have been forced and the army would have been cut to pieces.'

Two Bell topped officers shakos and cross belt plate c1830.

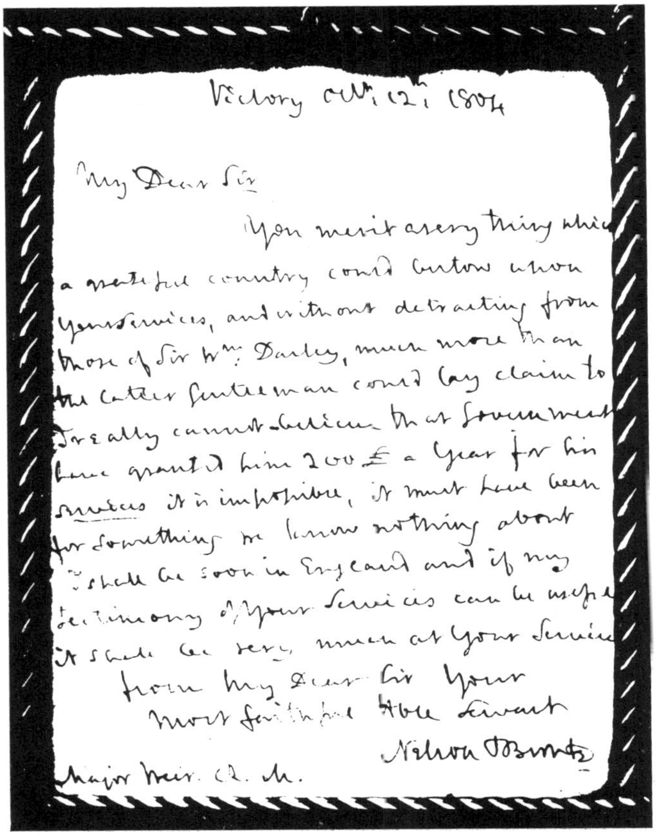

Victory Octr 12th 1804

My Dear Sir

You merit every thing which a grateful country could bestow upon Your Services, and without detracting from those of Sir Wm Darley, much more than the latter Gentleman could lay claim to I really cannot believe that Government have granted him 200£ a Year for his services it is impossible, it must have been for something we know nothing about I shall be soon in England and if my testimony of Your Services can be useful it shall be very much at Your Service from My Dear Sir Your most faithful Hble Servant

Nelson & Bronte

Major Weir. R. M.

A letter from Admiral Lord Nelson to Major James Weir, great-great-great-grandfather of the present Commandant General, Lt-Gen Sir Martin Garrod, thanking him for his services in raising the Malta Light Infantry. Major Weir was the senior Marine Officer at the Battle of the Nile, serving in HMS Audacious, and it is believed he later became the Governor of Malta.

CHAPTER 4

1840-1863
HIGH SUMMER OF EMPIRE

'It is a Corps which has never appeared on any occasion or under any circumstances without doing honour to itself or its country'

Marquess of Anglesey,
Speech at Portsmouth on 5 August 1841

One of the few known oil portraits of a Royal Marine Sergeant Major, George Dibben, c1847, which hangs in the Museum at Eastney, Southsea.

The following years were among the most diverse and interesting for the Corps as the long Victorian heyday of Empire waxed powerfully. Service at Sea was further reduced and their role with the Naval Brigades ashore predominated as the century progressed. The only break in the sequence of colonial wars and policing actions was the Crimean War and in particular the Baltic and Sebastapol operations. As for the Corps itself there were many significant changes during this period.

Action continued on the edges of the Ottoman Empire in the Mediterranean. In September 1840 the marines embarked in the fleet were landed, this time to help the Turks repel the unwelcome invasion of Mehemet Ali, Viceroy of Egypt. Here England, Austria, Russia and Prussia concluded a treaty with Turkey to prevent the take-over of Syria. A second Marine battalion was sent from England to reinforce those already ashore and together they formed a 1,500-strong Brigade of Marines under Commodore Napier. The Egyptians were quickly evicted from Beirut after the initial landings on 9 September. However a secondary attack, delivered against the castle of Djebail by 200 marines landed from Capt Martin's squadron, was driven off with the loss of 23 men. The Egyptians pulled out the next day and the fort was duly occupied.

Further down the coast the Lebanese town of Sidon was also attacked by a force of seamen and marines that day. It was defended by two strong castles which the fleet bombarded fiercely, and while the Turks took one fort, the British and Austrian marines secured the other. Further action by Napier ensured that the town surrendered the next day.

Acre, in the south, then became Ali's main stronghold in Syria and this too was put under siege. It had proved a tough nut to crack when the marines had defended it forty years earlier. It looked equally daunting now they were the attackers, with 147 guns mounted on its sturdy walls and a defending force of over 5,000. The British fleet arrived off the port on 2 November. As it happened there was not much fighting to be done as the preliminary bombardment by the fleet hit and exploded the main magazine, wiping out the best part of two regiments of defenders on the castle walls. This took the heart out of the Egyptians and all the marines had to do was to supply a small landing party to help the Turkish occupation force bring the insurrection to a speedy conclusion.

One wry comment was made in the aftermath of this little war:

Two Royal Marine Artillery Gunners in shakos, blue coatees and white trousers, c1854, at the time of the award of Crimean medals. In these early days of photography, Queen Victoria arranged for photographs of representatives of all regiments involved in the campaign to be taken for her collection.

Royal Collection, Windsor Castle

The British commander-in-chief and the fleet were voted the thanks of Parliament for their brilliant success at Acre, Commodore Napier and Capt Walker received the ribbon of the Bath, all the captains were made CB, 10 commanders, 23 lieutenants and 50 mates were promoted; but as usual the marines were passed over in the distribution of rewards.

The first China War came about as a result of a series of provocative steps taken by the Chinese Emperor against British trade, abrogating the East India Company's charter of 1833 giving them free trade rights. Four years later the Imperial Government reneged on this and Lord Napier's efforts to bring about a settlement were treated with contempt. British frigates on one of their lawful visits to Whampoa were fired upon by Chinese guns on Chuepee and Tycocktoo Islands, but they refused to reply. Turning the other cheek was naturally interpreted by the Chinese as weakness and provoked more incidents. The larger Bogue forts and the Bocca Tigris joined in, whereupon the British ships responded. This brought an abrupt end to the incidents and a treaty was agreed. However, it was soon broken when a fleet of junks attacked a British squadron at Chuenpee in November 1839, being roundly defeated for their pains.

On 5 July 1840 sterner action was taken against the batteries at Chusan to open up the Canton river once more. A party of troops was put ashore on the main island and the Mandarin forts were bombarded. Chusan soon fell and a few days later Amoy followed. The following January the fortresses of Chuenpee and Tycocktoo were stormed by troops and a party of seamen, while the Bogue forts hoisted a white flag before a shot was fired. Marines were in many of the ships that were engaged in these operations.

Another peace treaty was signed. Hong Kong was occupied by the British in January 1841 and formally ceded to Britain by the Treaty of Nankin the following year. However, almost at once the Bogue forts started firing again on British merchant ships, the *Nemesis* being attacked that month. This provoked further retaliation and on 25 February 1841 these forts were attacked once more. The one on Anunghoy island mounted 172 guns, and that on North Wantung 167 guns. Both were bombarded by a squadron under Sir Henry Senhouse until they succumbed. The marines and sailors were then landed and stormed both forts, carrying them after a short fight which resulted in 500 Chinese dead and 2,500 prisoners. The British then turned their attentions to Whampoa, where a battery and forty junks were captured by Capt Herbert's squadron.

On 23 May the Chinese fired on a flag of truce and hostilities were resumed. 380 marines under Capt

A recruiting poster c1850 which offers 'the bringers of good recruits - 3 guineas'.

Samuel Ellis participated and attacked the northern fortification of Canton, which was duly occupied. Maj-Gen Sir Hugh Gough had at his disposal a force of 2,000 troops and the marines were part of the Naval Brigade which stormed and forced the city gates. The city was then ransomed for six million dollars, but the Peking government remained defiant. In 1842 Admiral Parker decided to attack Amoy island, whose powerful defences had been strengthened still further. On 25 August the usual heavy British bombardment silenced the defenders' guns and the landing by sailors and marines quickly achieved the capture of those batteries on the seaward side of the city. Marines also participated in the occupation of the batteries on Kalungsoo Island and again at Chusan on 1 October. The Naval Brigade under Capt Herbert moved on to Chinae on the mainland, where more forts were bombarded and stormed.

The whole campaign continued up river with the fall of Nogpo. It was then decided to switch to the Yangste Kiang river, and to take the great city of Nanking in the hope that it would bring the Emperor to the negotiating table. Chapoo was taken and the batteries at the mouth of the river captured, Shanghai falling in the process. On 6 July Woosung was taken and on the 21st Chin-Kiang-Foo. Here the Naval Brigade, which included over 200 Royal Marines, particularly distinguished itself, and a veteran Royal Marine officer, Major James Uniacke, was amongst those who died of sunstroke during the battle. The Chinese finally sought peace on 29 August. The Admiral commented that it was with

> ...renewed pleasure that he again reports the zeal and gallantry evinced by every officer and man of the Royal and Indian Navies and the Royal Marines under his command...

On the other side of the globe, South America caused a diversion when the Argentinian dictator Rosas closed the Parana river to commerce. In November 1845, a combined Anglo-French squadron under Capt Hotham RN was despatched to teach him otherwise, and they attempted to force the floating barrier of 24 boats and a 22 heavy gun battery which was located by his forces at Obligado, 100 miles from the sea. Once the chain boom had been daringly forced, a strong naval bombardment was followed by the landing of Capts Hotham and Sullivan with 180 seamen and 145 marines, who stormed the upper fort and cleared the barrier.

Piracy in the Far East kept the Marines busy and in 1845 Capt Talbot RN led a force of 350 seamen and marines in 27 boats up the Songibesar river, in the Labuan area of Borneo, to attack a pirate stronghold. For the loss of 21 killed and wounded this nest was completely wiped out. The marines were to score another notable victory up the same river over a century later. The Commander-in-Chief Sir Thomas Cochrane personally led another attack in 1846, with over 600 men, against the Sultan of Brunei. They stormed five batteries of guns at Pulo and burnt down the capital. In March 1853 Pegu was taken under siege but General Godwin's column was hard pressed to raise it. A strong force of marines and sailors was landed to provide invaluable assistance.

In Burma the notorious *dacoit* Myatoon was causing havoc and on 4 February 1854, Royal Marines were part of a 225-strong force under Capt Loch RN which, with the 6th Bengal Native Infantry, attacked the fortified positions at Donabew. This early attempt at jungle warfare ended in disaster, the column blundering into the main Burmese defences and receiving a bloody nose. The situation was retrieved by a second column which included the 80th of Foot.

A change from skirmishing of this kind came with the war against Russia, which started on 28 March 1854. Fought (on the British side anyway) mainly to preserve the Dardenelles from Russian encroachment, the Crimean War provided two main theatres of combat for the Royal Marines; the Baltic, where a British fleet was despatched with high hopes, but achieved very little; and the Crimea itself, where the long and costly siege of Sebastapol saw the Royal Marine Artillery come to the fore. In both campaigns members of the Corps were awarded the recently created medal for supreme bravery, the Victoria Cross.

Uniforms of the Royal Marines Light Infantry c1855 from a painting by H de Daubra.

Although Vice-Admiral Sir Charles Napier's handling of the Baltic fleet left a great deal to be desired, as his strategy was restricted by his unnecessary awe of the Russian batteries, some action was taken in the summer - particularly the raid on Bomarsund in the Aland Islands in August. A combined Anglo-French fleet landed troops to reduce the fortresses there. The French went ashore on the south-west side of the main fort, while the British, under Brig-Gen Jones landed close to the northern fort with 700 marines and seamen.

Siege batteries were put in place and together with a heavy bombardment from the ships these guns helped to reduce the main citadel, which surrendered on 16 August. Further action was then directed against the battery on Presto Island, which also fell. Little else was done by the Baltic fleet that year.

Further north in the White Sea a more energetic approach was taken by a squadron under Capt Lyons RN with *Brisk* and *Miranda*, while in the Pacific another squadron commanded by Capt Nicholson RN with the *President, Picque* and *Virago* landed troops to take the Russian 22 gun batteries at Petropaulovski in Kamtschatka, which were protecting two of their own ships. Capt Burridge RN of the *President* led ashore a 700-strong party and captured the principal fort, but they were counter-attacked and thrown back with heavy loss.

Operations were resumed in the Baltic the following year, this time under Admiral Dundas. Initially small raids against Russian strong points were conducted, including one at Viborg, where a Royal Marine Artillery officer, Lt George D Dowell won the newly-instituted Victoria Cross. He was aboard the *Ruby* while his own boat was replenishing her supply of rockets, when he witnessed an explosion in one of the *Arrogant's* rocket boats close by. Without hesitation Dowell joined the rescue party aboard one of the boats lowered by the *Ruby* to go to their assistance. Although under heavy enemy fire both from muskets and grape-shot, he and three volunteers rowed over to the sinking rocket vessel. He not only saved the lives of three of the crew, but also brought the cutter safely away, still under fire.

On 10 August a larger attack on the fortress islands of the port of Sveaborg, which the Royal Marines supported with sixteen mortar boats and numerous gunboats, was attempted. Covering fire for the Russian defenders was given by the powerful fortifications of Helsingfors just across the bay. Capt J M Wemyss RMA was in command of the mortar boats, which anchored within 4,000 yards of the batteries and opened up with some accurate salvoes of their huge shells at the rate of thirty an hour. The fortress suffered severely from this barrage: huge fires were started in the town and large explosions were heard. The attack resumed next day a quarter of a mile closer in and the Vargon magazines were detonated. Having accomplished this, the Allied fleet withdrew.

But it was in the Black Sea, and particularly the Crimea, that the major involvement of the Corps was to be found. The first landings of the campaign were at Eupatoria when 400 marines were put ashore. Once it had become clear that the Russian fleet would not fight at sea, but had been scuttled at Sebastapol, the Royal Marines of the fleet were also put ashore to join the siege. Altogether 1,200 were landed along with the RMA batteries, the latter joining in the siege while the former were formed into a two-battalion brigade. One battalion, of about 1,000 men was commanded by Lt-Col Thomas Hurdle, and with a further 200 under Capt Hayes Marriott from the battleship *Agamemnon*, plus two RMA batteries with 26 guns, took over the entrenchments on the heights overlooking the main British supply port of Balaclava to the South. This position became known as the 'Marine Heights'. On 20 October, the Russians attacked but were thrown back and the Marine gunners 'made most excellent practice.'

Another Russian thrust on 5 November resulted in the famous battle of Inkerman. 300 Royal Marines under Capt W F Hopkins RMLI were in the line at the time and fought valiantly. Here it was that a small party under Sgt Richards was sent to flush enemy snipers from some caves in a hill and were surprised by a greatly superior number of Russians at close quarters. Cpl John Prettyjohn, a Cornishman, placed himself in the most exposed position and quickly shot four of the leading enemy soldiers outright. Then, being out of ammunition, he resorted to physical assault and literally threw the next Russian to close with him over the heads of those following. He continued with a fusillade of stones, driving the astonished enemy before him.

The siege of Sebastopol was a long drawn-out affair, and the dark winter months saw sickness strike down many a veteran and reduce the ranks of all troops considerably. Fresh drafts from England were full of young boys who quickly succumbed in their turn, and those who recovered had to be re-embarked for their own safety and survival.

In May of the following year raids were mounted against towns and ports on the Kerch peninsular, where the Russians had vast stocks of supplies and ammunition. A British squadron roamed the Sea of Azoz causing wholesale destruction, landing parties at Genitchesk and Taganrog as well as Mariopol and Gheisk. Back at Sebastapol in June came the first assault on the Redan, which was beaten off with grievous loss. The men of the RMA were still in the supporting batteries of the Right Attack and it was here on the 5th of that month that Bombardier Thomas Wilkinson performed heroically when his battery defences were smashed. Risking himself recklessly, he led the rebuilding of the earthworks despite a hail of shot and shell. He, too, received the Victoria Cross for his actions on that day, and for work done on numerous previous occasions.

On 2 February 1855 Queen Victoria announced that the Corps of Royal Marines should in future be designated a 'Light Corps' in recognition for their services in the Crimea to date. The announcements of the awards came some time after the actual incidents with the initiation of the award of the Victoria Cross by her Majesty. Sgt W Turner described how the colonel paraded his battalion and told them they all deserved a VC apiece, but he had only one to give out. Some men told him to take it for

Cpl John Prettyjohn wins his VC against the Russians at the Battle of Inkerman in 1855. Details of the citation are at Appendix C. From a sketch by Van Huen

himself for he had saved all of their lives when he had ordered them to lie down under heavy fire. But he refused saying 'No lads, it's for one of you; which shall it be, Prettyjohn or Jordan?' The response was for the former and the colonel awarded it thus along with the twenty pounds a year annuity that went with it to Prettyjohn '...for he is in his 21st year of service.'

With the taking of Malenkoff by the French, and the final occupation of the Redan following another successful assault on 7 September by the British, the Russians accepted the inevitable. The blowing up of the remaining magazines by the defenders followed and with their humiliating withdrawal across the boat bridge the siege came to its long and bloody conclusion. There were a few final operations. On 14 October Kinburn, a fortress town to the east of Odessa at the mouth of the Dnieper river, was bombarded and the marines landed as a storming party. Another attack was made on the mouth of the River Bug. Before the second winter of the war the Marines were withdrawn back aboard ship and saw no further action before the termination of hostilities in March 1856.

Little time elapsed between the end of the Crimean War and the ghastly massacres of the Indian Mutiny which began in May 1857. Although the Royal Marines presence was small it was often vital. The Provisional Battalion, which had been sent from England to China, was hastily diverted to Calcutta where the bulk of the Royal Marines under Lt-Col Thomas Lemon RMLI garrisoned Fort William. However, about 150 marines served with great distinction in the Naval Brigade under the energetic Capt William Peel RN, and saw hard fighting at Lucknow and elsewhere. Peel's Naval Brigade of some 403 marines and seamen with heavy guns, landed from the *Pearl* and *Shannon*. They set off up river on 18 August arriving at Allahabad, where they formed the garrison and were later reinforced by a further 120 seamen under Lt Vaughan RN. On 1 November the marines fought an action at Kajwa, while at Lucknow they supported the big 8-inch guns of the *Shannon* which contributed so much to the suppression of the strong-points of the

HM Queen Victoria about to present the Victoria Cross to Lt George Dare Dowell and Bdr Thomas Wilkinson at a parade in Hyde Park in 1857. This was the first such presentation. Details of their awards and citations can be found at Appendix C.

Painting from Royal Collection

Sepoy mutineers of Dilkhoosha, the Martinère and Shah Nujeef.

The Royal Marines also fought at Futtegurh in December, approaching the hanging bridge by boat and stopping the mutineers from dismantling it. At the siege of Cawnpore the following January Peel died of smallpox, but his gallant Brigade continued to give the army close heavy-fire support and took part in ten pitched battles with the rebels. Lt Edward Pym RMLI led another party of 150 marines in the battle of Sebanpore on Boxing Day 1857.

We have mentioned the interruption in the journey of the Provisional Battalion to China caused by the uprising in India. Let us now return and examine the work of the Royal Marines in subsequent China Wars. These were, in truth, continuations of the first, and were brought about by the arbitrary and treacherous breaking of signed treaties by the haughty Mandarins of Peking, who regarded the British as barbarians and refused to treat any agreements made with them other than as convenient breaks in hostilities allowing them to re-group and re-arm. On the British part there was reluctance at home to commit more than the bare minimum of forces required to effect a result. So the wars became a series of individual campaigns that dragged on for five years and were not finally settled until 1860.

Once more the war was sparked off by a series of incidents on the Pearl River, which led to the British occupation of Canton. The seizure of the British registered vessel *Arrow* by the Chinese authorities at Canton was the initial spark and the British Commander-in-Chief, Sir Michael Seymour, responded vigorously with an attack on the Barrier Forts on 23 October 1856. Imperial Commissioner Yeh remained abusive and defiant and so Seymour moved against the city itself, destroying a fleet of war junks, taking the 200-gun Bogue Forts and occupying the residential parts of Canton. On 29 October, 200 marines led by Capt Penrose attacked the strong city walls and managed to gain a lodgement inside the great gates. But they were considerably outnumbered and had to withdraw. As the forts were taken, one by one, the Chinese burnt British factories and so parts of the city were put to the torch in retaliation. Several marines who became cut off from their comrades suffered the true barbarity of the Chinese and were beheaded. Part of a poem written by Lt S S Sugden RN records this event:

> The Emperor sat on his Peacock Throne,
> An English soldier stood there alone...
> "Uncover at once" a mandarin said
> As he smote the cap from off his head...
> "Tell him that I, as a Royal Marine,
> Always stand covered before my Queen.
> Not to uncover before him I swore,
> No. Never will I disgrace my Corps"
> A mandarin yelled "Kow tow, kow tow,"
> "No" said the captive "I don't know how.
> I'm only a private, a Royal Marine,
> Bend two knees to God and one to my Queen.
> I'll give him a soldier's salute or bow;
> But not more than that; I'll never kow tow"
> "Drag him away" the Emperor said;
> "Make him kow tow and bring me his head"...
> A hero died. He has taught us how
> To die like men, but to never kow tow.

While awaiting the arrival of the Provisional Battalion, further operations were begun in May 1857 when another fleet of war junks was destroyed at Escape Creek. In June, 250 marines were landed and occupied a Chinese fort as a diversion which enabled the sailors to engage the enemy in the larger battle of Fatshan Creek, where an even bigger Chinese battle fleet was destroyed. The Chinese garrison of Fatshan patrolled out were met by the full force of more marines landed from the fleet, who soundly thrashed them. The outbreak of the Mutiny, as we have seen, brought about an enforced lull for a time while the Provisional Battalion was diverted to India.

In September 1857 Lord Elgin returned from Calcutta and the surrender of Canton was requested. The 1st and 2nd Battalions of Royal Marines straight from England, along with 100 men of the RMA under Major Joseph Travers RMLI, were again joined by the 300-strong Provisional Battalion from Calcutta under Lt-Col Thomas Lemon, with Prettyjohn now a Sgt-Major. Yeh again remained aloof, and so, on Christmas Day, after a heavy bombardment from mortar ships, rocket vessels and gunboats on the defending forts and installation, a landing party of Royal Marines, sailors and 800 French soldiers went ashore. For the loss of 130 men the great city of three million people was taken after a few days fighting. Still no answer had been obtained from Peking, so the Allies sailed to the mouth of the Pei-Ho river and demanded the surrender of the great Taku forts that guarded its mouth. A mixed landing party was put ashore on 20 May and they overran the seemingly impregnable fortifications, which stretched for a mile along both mud flats and banks of the great river very quickly. This persuaded the Emperor to sign yet another treaty at Tientsin.

Once again this seems merely to have been an excuse by the Chinese for a breather while they re-occupied and greatly strengthened the Taku forts, laying long chains and barriers across the river mouth and mounting extra heavy guns. Meanwhile Admiral Sir James Hope had replaced Admiral Seymour and when the British and French envoys, Lord Elgin and Baron Gros, were refused entry to ratify the Tientsin Treaty, fresh Allied preparations were started to force the issue.

Col Lemon's Provisional Battalion, with a second battalion raised from marines from the fleet, were held ready and the admiral ordered a reconnaissance of the forts. It was found that many extra guns had been positioned and two ditches dug, reinforced by rings of sharply pointed wooden stakes (panjis). Admiral Hope decided on a frontal assault across the mudflats against the 58-gun fortifications on the south bank, but Col Lemon protested strongly that this would be suicidal. And so it proved.

On 25 June the attack was nonetheless launched. The gunboats were raked by the guns from both banks, and when the 400 marines struggled ashore that evening they sank up to their knees in glutinous mud. Under heavy fire they forced onwards, men falling thick and fast, among them Lemon and two naval commanders. Only 150 men reached the first ditch and 50 the second, which proved insurmountable. A withdrawal was ordered and the survivors and wounded somehow staggered back to the boats and re-embarked as night fell. It had been a miserable disaster with 80 killed and 350 wounded. Only the heroism of men like Lt Harry Lewis Evans RMLI, who three times went back under fire to rescue wounded from the morass, gave some uplift to a miserable day. Probably because this had been a defeat such heroism, which many said deserved a VC, went unrewarded. Admiral Hope could do nothing but withdraw his fleet to Ningpoo and await his chance.

Revenge was not to come until the following year when the surviving Marines were amalgamated into one battalion which, in 1860, was sent to Chusan and Shanghai. Meanwhile, a fresh army was being assembled under Sir Hope Grant with French support. A full scale land campaign started which after several major battles re-took the Taku forts and pushed on up the Pei-Ho, through Tientsin, to the walls of Peking itself. Although the marines played

An officer of the Royal Marine Artillery in 1844 standing in front of the Royal Marines Barracks at Woolwich. This was known as the 'Court Division' owing to its proximity to London.

From the painting by Charles Stadden

Right:
A group of officers of Woolwich Division c1860. The Royal Marines Light Infantry formed the Fourth Division in 1805 and after various moves, the Admiralty erected Cambridge Barracks for them which they first occupied in 1848. The Woolwich Division was abolished when the Corps strength was reduced in 1869.

Forton Barracks, Gosport, the home of Portsmouth Division of the Royal Marines Light Infantry, which they first occupied on 29 March 1848. They remained there until they amalgamated with the Royal Marine Artillery on 22 June 1923, with their last parade at Forton on 1 August. HM King George V, as Col-in-Chief said 'It is with great regret that, in consequence of the reduction in numbers, and the necessary financial economies necessitating the abolition of one of the historic Divisions, I have concurred in the amalgamation of the Royal Marine Artillery and the Royal Marine Light Infantry'.

little part in these actions, a 400-strong detachment of Royal Marines was present at the end and were among the first to enter the city. Two companies remained at Shanghai after the Tientsin Treaty was finally ratified in November. Strangely enough the Marines' next actions in China were on the side of the Emperor, for they were part of a force of seamen and marines which helped Col H Ward defeat the Taiping Rebel army at Woosung. In all, some 232 officers and men had died in China in this campaign and 222 more had been wounded, many of them severely.

On a more formal note, on 1 November 1859 the Royal Marine Artillery was formed into a separate 3,000-strong Division and their Headquarters were to be a brand new barracks built at Eastney, Southsea, which was finally opened in 1867. The separation of the Royal Marine Artillery and the Royal Marines (Light Infantry), as it had been since 1855, was formalised under the provisions of an Order in Council dated 21 March 1862. However, in 1869 the 'Court' Division at the 'Red' Barracks at Woolwich was closed as an economy measure. That matters had not changed much since the days of the Stuarts was shown by the harshness by which this reduction was conducted. Officers were placed on half pay and men turned away at the gates of the barracks. The true face of much-vaunted 'Victorian values' was clear for all to see.

Royal Marines continued to serve around the world, but not always seeing the action they joined for. Like sea-time through the centuries, much of it could be boring and uninteresting, as was the food. There were occasional troubles on the mess decks and on runs ashore in foreign ports. One such, recorded for posterity in the *Guinness Book of Records*, is that of Marine John Dalliger. He served aboard the *Leven* in the Yangste, and on 13 July 1860 was the last person to be hung from the yard-arm after being found guilty of two attempted murders.

In the Far East similar resentment to that encountered in China by nationalistic forces against foreign traders penetrating their land was also manifesting itself in Japan. The warrior cast, the dreaded *Samurai*, were stirring up trouble. An attack was made on the British Legation at Yedo in July 1861 by a group of these extremists known as the *Ronins*. Several of the guards were murdered and in June 1862 a British merchant was killed by these young fanatics. An Anglo-Dutch fleet bombarded the forts at Kagoshima on 11 August 1863 as a show of strength, but the troubles began to spread. It was thought that the Legation at Yokohama was also under grave threat of attack. To reinforce its guard an RMLI battalion was assembled at Plymouth and embarked aboard a former battleship which had its main gun battery deck removed to act as a troop transport. This ship, the *Conqueror* duly arrived at Yokohama and disembarked the marines: in the event, this show of force was sufficient to quench the ardour of the majority of the Japanese rebels.

However, the RMLI were to see a limited form of action during their stay in Japan. Prince Chosiu threw in his lot with the *Ronins* and effected a blockade of the Inland Sea from his base in the province of Simonoseki. An Allied force consisting of English, French, Dutch and American ships was therefore despatched in October 1864 to break this blockade, and the RMLI battalion sailed with them aboard the *Conqueror*. This battalion, commanded by Col P C Penrose RMLI, and another made up from marines of the fleet under Col C W Adair RMLI was formed into a Royal Marine Brigade under Col W G Suther to which was added some Dutch marines. They landed and advanced inland but met no resistance, Prince Chosiu recognising the futility of opposing such a powerful force. During their withdrawal after negotiations, some of the Prince's more resolute followers opened fire on the marines from a hidden stockade in a forest ravine. At once a storming party under the command of Major W H Poyntz RMLI attacked the strong point, clearing it with great panache. Inside they found several warriors clad in the traditional armour suits, both chain and plate. Those they did not kill fled into the forest, thus ending the affair. The battalion returned to Yokohama where they remained until September 1865, when they were brought home. Another Royal Marine battalion went out to Japan five years later. They were commanded by Lt-Col E Richards and remained for five years, their duties being largely ceremonial as the Japanese had curbed the activities of the *Samurai*. The Japanese were much impressed by the bearing and discipline of the RMLI during their stay and profited much from their help in the modernisation of their own armed forces, which proceeded apace during the last quarter of the century. The battalion was accorded the singular honour of being invited to parade before the Mikado before they embarked for home. It was reported that

> ...as the battalion marched down - drums beating and Colours flying - to the pier, every man, woman and child, Japanese and foreigner alike turned out to bid them farewell. It was long before the settlement recovered from the loss of so considerable and genial an element in its society.

Col W G Suther and a group of the officers of his battalion in Japan early in 1864.

A Royal Marines Battalion under Col W G Suther, along with bluejackets from the Fleet stormed the Japanese stockade at Simonoseki on 6 September 1864. They destroyed the batteries and spiked the guns. Inquisitive sailors and marines are seen viewing one of the enemy batteries. (Below right)

A group of Royal Marines cadets, seniors on the right, juniors on the left taken in September 1862 from an album of Cadet W P Wright (pictured far right seated). They trained aboard the wooden hulk HMS Excellent *moored in Portsmouth Harbour. All but one of these cadets were RMLI, and they spent a year under training before being commissioned as Second Lieutenants.*

S.Y. Alston
and
W. P. Wright
Woolwich
19th June

CHAPTER 5

1863-1885
FAR FLUNG CORNERS

'E was scrapin' the paint from off 'er plates, an' I sez to 'im "'Oo are you?"
Sez 'e, " I'm a Jolly - 'Er Majesty's Jolly - soldier an' sailor too!"

Rudyard Kipling. *The Seven Seas*. 1892.

From the 1860s onwards Great Britain reached the peak of her power and influence. Although the Royal Marines continued to serve at sea, the fleets changed beyond all recognition during this period. Wooden walls and broadsides still predominated in 1863 but the arrival of the *Warrior*, the first of the iron-hulled frigates, began a revolution of enormous potential. Masts, spars and sails gave way to the pounding of the reciprocating engine and huge funnels; muzzle loading cannons on trucks firing on the broadside were replaced by enormous barrels of up to 18" calibre housed in rotating turrets discharging huge, rifled shells from 20 miles and more. As the ships, their armament and equipment changed, so too did the men.

Many felt that the Royal Marines, probably the most versatile of all the military and naval formations in the Empire, were too restricted and hidebound to compete with these changes. The Royal Marine light infantryman of 1885 was basically the same staunch and steady rifleman he had been in 1863. His uniform had altered, his weapons had been improved but he was still, first and foremost, an individual subordinated to a line or square formation, in which he would stand and fire accurate volleys until he was ordered to advance with fixed bayonet and complete the job with cold steel. True, he might have a rocket battery to assist him, and his well-aimed volleys might be supplemented by a Gardiner or Gatling machine-gun, but the ultimate result of a land battle would depend on his steadfastness and discipline rather than on his individuality.

Lt S V Alston and 2nd Lt W P Wright, both of the Woolwich Division in 1863, wearing undress uniform of frock coat with crimson sash over the left shoulder as was the custom with officers of this period. This informal photograph is copied from a typically ornate layout in one of 2nd Lt Wright's personal albums held in the Museum.

A Private of the Royal Marine Light Infantry, 1865. Charles Stadden's painting depicts him standing near the Main Gate of Forton Barracks, Gosport, first occupied by the Portsmouth Division in 1848.

His comrade in the Royal Marine Artillery was to feel even more alienated from the progress of the second half of the 19th century. This may sound strange, for it was a technological age in which the gunner above all should have felt supreme. But for the marine gunner afloat it was a period of frustration as less and less did he feel he had a place in the scheme of things at sea, until with Admiral Jackie Fisher his very existence was questioned and he was regarded like a dinosaur who had outlived his usefulness. This was wrong, as we can now plainly see in retrospect, for the Marines were already ahead of their time. In this period drill in white canvas tunics at a machine-gun aboard a battleship, or practice firing one of the great Armstrong weapons, was poor compensation for a precise role within the Navy's domain.

Nor did the prospects of promotion or a settled future become any more certain; in fact, the Corps continued to be shamefully neglected in both aspects

In November 1859 Capt W A Delacombe, with two other Captains, five Subalterns and 200 men were sent to British Columbia as garrison troops for San Juan Island. He is pictured with his family in 1866, and remained there until 1872. The old block houses on the islands are still standing and provide a tourist attraction.

during Victoria's heyday, as they had been a century before. Too many Royal Marines captains and subalterns were grey-haired before they achieved higher rank in the service. The fact that, ceremonial and drill apart, officers had little gainful employment afloat and, perhaps, more galling, little or no control over their men who were ordered about by young naval officers scarcely out of public school, produced a mood of growing cynicism. Any ambitious or specially competent young officer could only enhance his career by luck or transfer to the army.

Yet there were opportunities for a fighting man to excel in the colonial wars and expeditions, which increased as the century grew older and Britain's dominion over the world continued to expand. In New Zealand, for example, a small number of marines had earlier participated in the Maori Wars, being present at Heke's pah at Okaihu in 1845 and Ruapekapeka in 1846. In June 1860 they fought at Putitakaneri and in February 1864 at the storming of Ta Awamuta. A *pah* was a stockade, which the Maoris were adept at building. Even so they should have posed little problems for any civilized force who had artillery to demolish them before the infantry assaulted. But time and time again vainglorious charges were made against them and such rashness received its due reward with heavy losses and frustrating reverses. Typical of such futile assaults was that on the Rangiriri pah by 200 seamen and 1,100 soldiers during the first Taranaki War, which broke out in 1860. No less than four frontal attacks were made before any attempt at breaching the stockade with artillery fire was made. Consequently 35 young men were killed and 85 wounded to no purpose. The Maori defenders numbered only 183, and must have thought this white man crazy not to use his big guns to settle the issue at close range.

There were other strange diversions too. The advent of President Miramon's rule in Mexico was brought about by a limited intervention by Britain

The RM garrison of San Juan Island. Lying close to Vancouver, the island was in dispute in 1855 in what came to be known as the 'Pig War' and a small garrison each from the USA and Canada occupied part of it. In 1859 the Americans reinforced their garrison and a Royal Marines detachment commanded by Capt George Bazalgette was sent there. The two garrisons lived in harmony for 12 years when the ownership was arbitrated upon by the German Emperor in favour of the United States.

into that country in 1861. This included a detachment of mounted marines, thus giving flesh to the legend of 'Horse Marines' which came about when the 17th Light Dragoons had embarked aboard the frigate *Hermione* as a marine detachment on the West Indies Station in 1795.

Another incident which involved the Royal Marines was the so-called 'Pig-War', a dispute that lasted on and off from 1860 until it was finally resolved in 1872. The area involved was Friday Harbour in the San Juan islands of British Columbia. Originally discovered by the Perez in 1774, and visited by Cook a few years later, the settlement of Columbia on the north-west shores of America had become a British colony in August 1858. San Juan island, in the Strait of Juan Foca which separated Vancouver Island from Washington State, was strategically important in marking the boundary between British and American influence, and both parties laid claim to it. Both sides therefore maintained troops on the island during this period. Between 1866 and 1867 Capt T Delacombe commanded a Royal Marines garrison at the camp of San Juan to ensure a British presence in the face of American encroachment and claims. Luckily this very sensitive situation never led to outright conflict, and was finally resolved when Vancouver Island was incorporated into it in 1866 and subsequently annexed to Canada in 1871.

Closer to home, the unending feuds in Ireland had led to the despatch of marines in a peace-keeping role during the attempted rising in 1848. Twenty years later, bloodshed and unrest by the Catholics against the Protestant minority called for a calming influence and a Royal Marines detachment again saw service in that troubled land. Such duties were unpopular for, as always, the poor 'Jolly' found himself in the middle

A group of RMLI officers at Forton Barracks, Gosport c1867. They are wearing the last pattern shako with the light infantry green ball on the front, which was replaced by a blue cloth helmet in 1878. The single-breasted scarlet tunic had a 'slashed' cuff at this time and the wearing of white trousers in the summer at home was abolished in 1868.

of what, to the average outsider, was an incomprehensible religious and bigoted squabble. How little has changed today.

In West Africa a force of 110 Royal Marines from England reinforced by seamen and marines from the *Barracouta*, all under the command of Brevet Lt-Col Francis Festing RMA, was landed on the poisonous shores of the Gold Coast (now Ghana) when the fierce and barbaric Ashanti tribes invaded the coastal regions of the Fantee Protectorate from the dense jungle hinterland in 1873. They were based at the old Dutch castle of Elmina. Known as the 'White Man's Grave', this fever ridden stretch of swamp and jungle became the scene of vicious fighting when mad King Koffee ordered his 12,000-strong army of tribesmen from Kumasi to claim the protectorate as his own. The small British force met a regiment of this army 2,000-strong at Kingstown and roundly defeated them, driving them back into the bush after heavy fighting.

A full-scale expedition was ordered at home led by General Sir Garnet Wolseley and this arrived in the Gold Coast in October of the same year. Meanwhile disease had reduced the ranks of Festing's little band and the rest were taken back on board for their own well-being. In November a force of 50 Royal Marines, holding the outpost of Abrakrampa were attacked by another column of 10,000 Ashanti warriors but managed to hold out for two days. A fresh RMLI battalion, sent out from home as reinforcements were kept aboard ship and not landed. Nevertheless, some 160 Royal Marines managed to be included in the column that set off into the dense rain forests on 14 October and inflicted a defeat on the natives in the bush.

There then followed the epic march on the evil capital and the termination of King Koffee's ambitions. A narrow path had to be hacked through the jungle, and resting places prepared for the troops along the way to keep them off the fever-ridden ground at night while quinine was administered liberally. The taunts that the 'White Man dare not enter the jungle' were firmly quashed, and the town of Essamen was reached after a steady, but exhausting, march. Another column consisting of a Naval Brigade under Capt William Hewett RN with 70 marines made a diversionary march across the River Prah on 20 December and Wolseley set off again with his main force towards Kumasi; but there were only a few marines there as Festing had been wounded and many of his men were sick.

Those in the Naval Brigade, however, remained to witness the final stages of the war. The enemy had retreated without a fight and so the column pressed on into the jungle, adopting a square formation for their defence as the scouts hacked their way on and gradually drew close to Kumasi. They were being continually sniped at from the thick foliage by the unseen enemy.

The Ashanti finally made a stand at Amquana where an ambush was turned by the Royal Marines who moved out on to the column's flank to clear the line of approach. The capital was finally reached and put to the torch before the column withdrew to the coast and the Ashanti gave no further trouble. Sir Garnet paid a tribute to this maritime comrades thus '...all have fought with the dashing courage for which seamen and marines are so celebrated.'

Royal Marines were present in Malaya during 1874/75 when, following skirmishes, the Sultans of Perak and Selangor signed treaties acknowledging British trading rights. They also took part in the punitive expedition to the island of Zanzibar in 1875, and the following year were present at incidents on the Niger Coast Protectorate until 1877. Nor was this the only part of Africa to see them during the ensuing years.

Little known today is the Transkei campaign which was fought in 1877, and yet Royal Marines were very much to the fore. In December a Naval Brigade of 200 seamen and marines with seven guns was put ashore from HMS *Active* at East London, South Africa, to take part in what is more commonly known as the Ninth Frontier War. The Commander-in-Chief was General Sir Arthur Cunningham and one column commanded by Col Glyn was sent out towards the Quora river where sailors of the Naval Brigade under Capt Wright RN with RMLI Lts J Craigie and B Hamilton were attached to this column. The Royal Marines formed part of a second column which pushed over and beyond the river into the bush, forming the hammer which drove the tribesmen onto the anvil. Here they were duly crushed in skirmishes on 13 and 29 January 1878 and largely routed at the Battle of Qaintana Mountain on 7 February. By 16 March most were back on board ship, their duty done. However a few, including the naval rocket battery, took part in the final battle of the war under General Thesiger, soon to be better known as Lord Chelmsford, when he finally crushed the Gaika tribesmen in the Peri bush.

Another incident took place in the Mediterranean at this time. The island of Cyprus had been ceded to Great Britain by the Anglo-Turkish Convention of 4 June 1878 and the Marines were landed in force to oversee the change of ownership. Back in South Africa, events were taking a much more serious turn with the advent of full-scale war in Zululand. On 19 November 1879 the *Active* once again landed her Royal Marines detachment with some seamen at Durban to assist the army in its new task. 172 men in all were landed under the overall command of Capt Campbell RN and they were equipped with one of the new Gatling machine-guns, two rocket tubes and two smaller guns. Their task was to hold an important crossing, Lower Drift, on the Tugela river which formed the boundary with King Cetshwayo's Zulu domain. They quickly built a fortified earthwork, which they named Fort Pearson, some 104 miles from the Zulu capital of Ulundi. They were reinforced by the Buffs and another 40 men landed from the *Tenedos* on 4 January. They found that they were not facing a few thousand Basuto tribesmen but a fully disciplined army of at least 40,000 men, feared the length and breadth of South Africa. Organised into *Impis* and fearless in battle they could march fast and long, relying on overwhelming numbers with which first to envelop, and then submerge, their enemies. At close range, with their short stabbing spears (*assegais*) and big shields, they were an enemy to be respected. Unfortunately Lord Chelmsford under-

The water tower at Eastney in 1868. Seated in the foreground outside the entrance to Eastney house is Colonel Commandant Sir George Schonberg and his family. This photograph was taken before the conversion of the tower to a Clock tower, but the other buildings remain much the same today.

estimated them completely, dividing his force into several columns instead of concentrating it and then splitting his main column again. The result was unmitigated disaster at Isandlwana on 22 January 1879.

The Royal Marines formed part of the southern column under Col C K Pearson which marched out of Fort Pearson on 12 January and constructed another fort, Fort Tenedos, on the Zulu side of the river, before moving on towards Eshowe where a forward base was established. Before they even got there the marines fought in the battle of Inyezane on the 22nd, where a force of over 5,000 Zulus was routed after much hard fighting. Eshowe was reached next day and quickly fortified. It was here that they heard the news of the disaster to the centre column. They remained surrounded but no attack came. Another naval brigade, under Capt Brackenbury RN, formed part of Lord Chelmsford's relief column which fought at the battle of Gingindlovu on 2 April.

There was no other action for the Royal Marines detachments although they formed part of a 833-strong Naval Brigade commanded by Capt Campbell RN which joined General Crealock's 1st Division during the second invasion. They only reached Fort Durnford where they were halted and returned to their ships, the final battle of Ulundi having been fought without their assistance. Incidentally, the Colonial Secretary at Pietmaritzburg at this time was Lt-Col Charles Bullen Hugh Mitchell, a former Royal Marines Officer.

A ceremonial parade of the Royal Marine Artillery at Eastney to celebrate the birthday of HM Queen Victoria on 24 May 1869, only two years after the barracks was occupied. The Band of the RMA are in the foreground. Years later dining halls and a main galley were built on part of this parade ground.

Ireland was the scene of yet further disturbances in 1880. Once again, detachments of the RMLI were stationed there in their now familiar role of targets for the local populace, withstanding the jeers, torments and stone-throwing of that section of the Catholic population who have seldom allowed reason to stand in the way of hatred.

One of the most outstanding contributions the Corps made to the cause of the Empire was seen during General Wolseley's brief, but devastatingly effective, Egyptian campaign of 1882, when the rebellion of the disgruntled Nationalist leader and Minister of War, Arabi Pasha, against the Khedive Tewfik, led to the massacre of European civilians at Alexandria which took place on the 11 June. The Khedive's Government was inert and the suzeraine power, the Porte, also remained indifferent. France, with whom Britain had hitherto acted in concert in Egypt, turned her back and so the Commander-in-Chief Mediterranean, Admiral Sir Frederick Seymour, was left to take matters into his own hands. The Egyptian army started work on port batteries and fortifications which were pressed ahead with a view to attacking ships of the fleet. Several ultimatums for this work to be halted were ignored, so the British fleet sailed out and bombarded Alexandria on 11 July.

This silenced the Egyptian batteries and the next day a force of seamen and Royal Marines,

A Royal Marine Artillery Colour Sergeant prior to 1881, showing him wearing three chevrons on the left sleeve and his Colour Sergeant's arm badge on the right arm as was the practice until 1881.

The Royal Marines Barracks on Ascension Island where the British flag was first raised early in 1815. There were Royal Marine Commandants from 1819-44 when the Admiralty took possession. On 27 October 1922 administration of the Island passed to the Colonial Office. A detachment of Royal Marines was maintained there continually during this time. Until 1922 babies born on Ascension Island were deemed to have been born at sea and registered in the parish of Wapping.

commanded by Capt H Fairfax RN, was landed to occupy the Ras-el-Tin area of the town in an attempt to prevent looting and murder. Despite this precaution much of the European quarter was burnt down. 700 marines with Gatling guns were eventually deployed in the town, not only to give protection to the Khedive, but also to prevent further violence in the streets. Finally, on the 17th, the Channel Squadron arrived with yet more reinforcements and eventually a mixed RMA and RMLI battalion under Col H B Tuson RMA was established as part of Maj-Gen Alison's occupation force.

On 2 August Admiral Sir William Hewett of the East Indies Squadron landed marines from his squadron at Suez without meeting any resistance. Three days later, Tuson's battalion was engaged with rebel forces at Melleha Junction. On 8 August a second mixed RMA/RMLI battalion arrived under Lt-Col S J Graham RMLI. All this was but a preliminary to the seizure of the whole canal which was carried out on 20 August from both the Port Said and Suez ends simultaneously. Three companies of marines were embarked in the gunboats *Dee* and *Ready* to help repossess the telegraph station at Kantara and take over all the dredgers and barges of

An early photograph of the main Gateway and Officers Quarters of the Royal Marine Barracks at Stonehouse, Plymouth, taken around 1880. These barracks were first occupied in 1783, though many alterations took place subsequently. Plymouth was known as the Third Grand Division.

the Canal Company which, due to the spleen of Ferdinand de Lesseps, was refusing to co-operate with the Khedive's wishes.

A detachment of seamen and marines, embarked in the gunboats *Mosquito* and *Seagull*, seized the docks and closed the lock gates. At Port Said Capt Fairfax organised the landing. The Egyptian sentries were surprised and rendered *hors de combat* by

The 30-strong band of Portsmouth Division RMLI under Bandmaster J F C Kreyer, a German, which accompanied HRH The Prince of Wales (later King Edward VII) in HMS Serapis *on a Royal Tour to India and Ceylon in 1875/76. In recognition of their service the band was awarded the Prince of Wales plume on 5 July 1876. On amalgamation in 1923 most of the musicians were transferred to the Depot band at Deal where they continued to wear the badge.*

Royal Collection

Lt-Col Tulloch and six marines who had landed from an open boat. At the same time, a company under Capt R P Coffin RMLI landed from the *Iris*. Equipped with a Gatling gun they took over the Egyptian barracks, while another company under Capt F M Eden RMLI, also with a Gatling gun, was put ashore from the *Monarch* to surround the same area. Outwitted, the rebels lay down their arms without a shot being fired. The only resistance met was at a village on the outskirts of the town, where a few stalwarts were quickly evicted by an advance covered by the ship's gunfire.

The Royal Marines ashore were now re-organised into an RMLI battalion under Graham and an RMA battalion under Tuson. The RMLI became part of Maj-Gen G Graham's 2nd Brigade. At Kassassin on 9 September, during the preliminary advance from Ismailia along the railway line and the Sweetwater Canal, a marine detachment with some captured mobile guns on a railway truck were heavily engaged and outnumbered, but held firm and eventually repulsed the Egyptian thrust.

Finally, on the night of 13 September, the long planned attack on Arabi's powerful fortified lines at

A Guard of Honour formed into part of a 'square' to repel a possible attack during the Royal Tour of India in 1875. Royal Collection

Tel-el-Kebir took place after a long night march. Col C Field, in 'Britain's Sea Soldiers' described how they were halted about 800 yards from the enemy positions and ordered to lie down and fix bayonets:

> Four companies were then extended as a firing line, followed by the remaining four as supports. We advanced by rushes, lying down to fire. The enemy's bullets were still falling thickly and the entrenchments and our own firing line blurred with smoke when the bugles - to the relief, I think, of everyone - sounded the "Charge". The whole line, every man burning to get at the Egyptians, rushed forward at the double with a continuous shout or roar rather than a cheer.

The attack delivered at bayonet point burst over the enemy holding the strong earthworks and trenches, which were lined with a large number of heavy guns, like a torrent. Only the Sudanese regiments stood and fought for a brief time, but even they soon joined the fleeting *felaheen* and the powerful defences were taken in a matter of minutes. By dawn, the whole Egyptian army had been routed and the desert was littered with the bodies of the fallen, blood staining their dirty white *djibbahs* and red fezes. The

A group of Royal Marines Officers at Chatham in 1881. They are wearing the blue patrol working jacket and blue cloth forage cap with RMLI cap badges. The Colonel Commandant (centre) wears gold lace round his cap. It is probable that the two civilian clothed officers were doctors.

Royal Marines lost Major H Strong RMLI, Capt J C Wardell RMLI and eleven men killed with 54 wounded. By 19 October, Cairo having fallen and the Khedive re-established, the marines marched out of the Egyptian capital to rejoin their ships.

Within a few short months the Egyptian-administered Sudan was aflame with war. A religious fanatic, proclaiming himself the Mahdi, preached that he had come to save his people from foreign oppression. He promised his followers immunity from death, or a world of pure pleasure in the after-life if they perished in battle; and he soon raised a massive following. Two ramshackle Egyptian armies with British officers sent to subdue them were slaughtered by the fierce Dervish warriors or 'Fuzzie-Wuzzies' as they were known. In 1883, General Gordon, sent out to Khartoum to evacuate the surviving Egyptian officials, was cut off and surrounded. While the

British government under Gladstone dithered and wrung its collective hands but refused to act, the revolt had spread to the Beja tribes of the Eastern Sudan, the area of the vast Nubian desert bordering on the Red Sea.

Here the port of Suakin was menaced by one of the Mahdi's local leaders, Osman Digna, known as 'Osman the Ugly', who was a slaver by trade and a religious fanatic by convenience. As Suakin bordered the route to India from the Suez Canal the British government was alert to the dangers, and Admiral Hewett of the East Indies Squadron was authorised to land a strong force of 500 marines to secure the town. Of these, some 140 officers and men were on their way home to England from China and Aden, while others came from the Mediterranean Fleet's ships. They were placed under the command of Major G H T Colwell RMLI of the battleship *Alexandra*. Meanwhile, General Graham was again sent out to Trinkitat with two strong brigades formed from regiments from England and India, to which a Royal Marines battalion was added to the 2nd Brigade. This brigade was transhipped from Suakin to Trinkitat, and on 28 February they marched out to relieve Tokar and made their way through baking heat to the mimosa-scrub desert towards El Teb, three miles away, reputedly Digna's base and the scene of an earlier Egyptian rout. They camped overnight at a zareba called Fort Baker and moved out in the boiling sun early next day to try to find Digna's Dervish army.

Find them they certainly did. The Mahdists were established on a hill to the north of the village, where they had dug trenches and installed some of the Krupps cannons they had captured in an earlier fight. They chained the Egyptian gunners to their weapons to serve them. The British infantry, formed into a huge rectangular 'square' 200 yards by 150 yards, marched stoically round the base of the hill, ignoring the hail of fire from the muskets and cannons of the wildly shooting tribesmen. The Royal Marines battalion, 380 men from both the RMA and RMLI, was on the left face of this square, which was the most exposed to enemy fire.

At about midday, after a 1,000 yards march when no shot was fired from the British side, the men were ordered to lay down while the field guns and Gardiner guns commenced a hot barrage against the Dervish cannon from a range of a little over half a mile. This silenced them quickly. The British troops fixed their straight 22" long three-channeled bayonets on to their Martini-Henry rifles. As the bugles sounded out the commands, they got up, wheeled to the left and started a steady advance up the hill towards the excited tribesmen.

This was too much for the Dervishes. They were not used to fighting in trenches or with cannons; their *fortè* was the wild unstoppable charge to close with their enemy, hacking and thrusting with their enormous Crusader swords and razor-edged stabbing spears. Hitherto none had resisted them, but now they found the British 'Tommy' was not to be overawed. In Col Field's words:

> The black fellows kept up a hot fire on the advancing British until they arrived within a couple of hundred yards of them, and then suddenly ceased firing and charged down on the square in their thousands, shouting, yelling and brandishing their formidable spears and swords. The stream of Martini bullets that rushed to meet them brought them down in their scores, but the survivors hurled themselves upon the bayonets, and steel clashed on steel. Many gallant soldiers bit the dust but the long triangular bayonet did deadly work, and when the Gardiner guns at the corners began to pump lead into the already decimated ranks of the enemy, they gave way. The British followed up the repulse with a rapid advance on the left of the enemy's position where they carried all before them.

The Dervishes then retreated to the roofless buildings of a ruined sugar-mill, whence they had to be winkled out with the bayonets once more; while a final stand was made at the El Teb Wells in the dip between the hill and the village. In all, the battle lasted for three hours and at the end 3,000 of the estimated 6,000 natives lay dead or dying. The Royal

Animal mascots have always been popular with servicemen. This collar, on display in the Royal Marines Museum, belonged to the dog 'Nell' who was buried at Eastney on 27 May 1885. She was a tanpariah bitch who attached herself to the Royal Marine Artillery during the Egyptian campaign of 1882. She saw action at Kafr-Ed-Danar and Kassassin where her master was killed. When the marines left Egypt, she swam out to the departing ship and, rescued, she spent her last years at Eastney.

Lt-Col Henry Tuson RMA wearing the distinctive blue uniform of an Artillery ADC to the Queen at Eastney in 1878, with the unusual gold embroidered frog-drop loop across the front of the tunic and across the cuff. His crimson and gold sash, worn over the left shoulder, was similar to that worn around the waist today by a Guards Officer in State Dress. The only item still worn today by an ADC to the Sovereign are the gold aigulettes. It was customary at this time for senior officers to provide special dress for their personal staff, as can be seen worn by the groom, Gunner Alfred Rimmington.

Marine casualties were four dead, a gunner from the RMA and three RMLI privates, and twenty wounded.

Another strong Dervish force soon gathered at Tamaii to the west. Calls upon them to submit received defiant answers, so General Graham decided to give them another lesson. Once more the brigade moved out with the Royal Marine battalion mustering a strength of 14 officers and 464 men divided into six companies. Another detachment of 100 RMA, commanded by Lt C G Brittan RMA, was left to garrison Fort Euryalus at Suakin. They were sniped at throughout the night in the zareba, but the advance started towards Tamaii on 13 March 1884 with the brigades in two separate squares in echelon to each other, the Marines being in the right-hand and foremost square with General Graham.

As they advanced they came to a deep ravine, or *kohr*, which formed a natural moat before the village. They came within 200 yards of it when the enemy made a series of attacks which were beaten off. These proved a lure. When the order to charge was given the British square became stretched and then broken because not every company charged at the same moment. This was the opportunity the enemy had been waiting for and hordes of them erupted, screaming from the *kohr*. They flung themselves upon the corners of the square and gained an entrance. They stabbed and hacked their way through, overrunning the naval Gardiner guns and forcing the marines to fall back, fighting as they went, for 800 yards.

It was a desperate moment, but soon the men were rallied. They formed into line and drove the Dervishes steadily back into the ravine, helped by the enfilading fire from the 1st Brigade's square which had come up in support, and also the work of the dismounted infantry on their flanks. Within ten minutes the ground was recovered, the guns recaptured and heavy fire was poured down upon the Dervishes huddled in the sides of the ravine. They began to fall further back, being steadily pursued until the village itself was reached and burnt. This costly little fight left three Royal Marines killed and fifteen wounded.

There was little other fighting, for the British

Five Hundred Royal Marines from the East Indies Squadron and Mediterranean Fleet were sent in February 1884 to reinforce Admiral Hewett's force at Suakin in Egypt. Some of the marines pose in front of their Medical Aid Tent.

government turned its face from mounting a relief column across the desert to Berber and recalled Graham and his regiments to England. Most of the Royal Marines followed, those who had been on their way home from eastern stations having priority and sailing on 29 March. The remaining 371 were retained to garrison Suakin before embarking in the liner *Utopia* on 14 April. In the meantime, on 10 February 1884, a fresh battalion had been sent out from home to Suakin. They constructed a proper defence perimeter around the town and built a series of makeshift forts and strongpoints which were named after the ships of the naval squadron.

Meanwhile the expedition to extricate Gordon was mounted up the Nile by General Wolseley with a detachment of RMLI serving as a company of the Camel Corps, a new variant on the 'Horse Marines' theme. After a hard fight at the Abu Klea wells, when the British square was broken for a second time by the Dervish charge, they were too late and Gordon was murdered as Khartoum fell. The expedition was therefore abandoned. One of the NCOs who served there, Sgt Eagles, was something of a poet, and he immortalised this incident when he wrote the following lines which were sung in the evenings to the accompaniment of a banjo:

When years ago I listed, lads,
To serve our gracious Queen,
The sergeant made me understand I was a Royal Marine.
He said sometimes they served in ships
And sometimes served on shore,
But never said I should wear spurs
And be in the Camel Corps.
I've rode in a ship, I've rode in a boat,
I've rode in a railway train,
I've rode in a coach and I've rode in a moke,
And hope to ride one again;
But I'm riding now an animal
A Marine never rode before,
Rigg'd up in spurs and pantaloons
As one of the Camel Corps.

In the Eastern Sudan events had repeated themselves, for General Graham was belatedly sent back there in 1885 with another brigade to cover the building of a railway from Suakin to Berber. It would have been easier the year before when, after two defeats, Osman the Ugly was on the run and deserted by many of his followers. But now he had had time to reorganise and so the job had to start from scratch.

Back again too came 153 officers and men of the Royal Marines from Suez to supplement those already there and join Graham's army on 9 March 1885.

Eleven days later, after an inspection by the general, the force marched out to give battle once more with the Royal Marines battalion being placed where they wanted to be, in front! The RMLI were part of Maj-Gen Sir John McNeil's 2nd Brigade, with the East Surreys, Berkshires and Shropshire Light Infantry. Their immediate objective were the wells at Hasheen, some eight miles west of Suakin. They marched through dense scrub, in fours, for several hours before reaching the head of the valley holding the Dervish camp, where they deployed into line. General Graham and his staff took up position on a small hill and the Surreys were held in reserve in a zareba atop an adjacent ridge. The Guards Brigade formed a square at the entrance to the valley below Hasheen Hill while the cavalry, the 5th Lancers and the Bengal Lancers were deployed to either flank.

The departure of the New South Wales contingent of Bluejackets and Royal Marines for the Soudan in 1885. They were given a tumultuous send off by the Australian crowds lining George Street, Sydney before embarking for the Middle East.

This left the RMLI and the Berkshires to attack the enemy, who were grouped in huge numbers on the surrounding high ground or lurked in the mimosa bushes that covered the hill quite thickly at this point.

The Berkshires then advanced in a line of skirmishers with half a battalion of Royal Marines in their rear covering the ground up the slopes against the natives, who replied with brisk but ill-aimed shots before being driven off the hill. As the *Army and Navy Illustrated* magazine reported:

> The second range of hills was held by the enemy in some force, but was cleared by the Royal Marines in splendid style, the rocks and precipitous slopes

being rapidly scaled and the Arabs driven off at the point of the bayonet.

Having established a post at the wells Graham then turned his attentions once more to Tamaii, where Digma was thought to have relocated his headquarters. Before embarking on this march, Graham considered it prudent to construct a zareba on the Tamaii road as an intermediate base and depot. Accordingly, on 22 March, he detached McNeil with the Berkshires, the Indian Brigade, the 5th Lancers and the RMLI to build it. It was to be organised in the form of three squares of cut mimosa bushes in echelon formation. By midday work was well under way in the scorching heat.

Lt F G Cotter RMLI described what happened next:

> The RMLI, having piled arms all round their square, were employed drawing water, rations, etc. Most had taken off their accoutrements and jackets. About 800 camels that had been unloaded inside, had been placed temporarily outside just south of the RM square. While thus employed, we were all suddenly startled by a roaring noise just like the sea would make in a squall. A cry of "They're upon us" and "Stand to your arms" was raised. This all came from the south. The cavalry scouts came galloping in immediately followed by a mass of Arabs. Those who came in in front of the Marines and Indians set to work hamstringing the unloading camels, the result being a general stampede of the camels and drivers - mostly Aden men - towards the Royal Marine square, sweeping away the feeble zareba, they darted through our square, carrying everything with them that was in their way, including about 50 marines, some Berkshire men, some sappers and a host of camp followers .....
> Hard after the camels came the enemy and about twenty of them penetrated our square, but out men, closing up, killed those who got in and drove back the remainder. The attack was well planned, but they did not understand the formation of three squares for mutual defence, and evidently reckoned on but one. They were puzzled.

Another officer who was there was to write in the *Globe & Laurel* thus:

> Some officer raised a great cheer, and this no doubt, greatly assisted to restore order again. Sgts Mitchell and Stanton (a volunteer from the Mediterranean Squadron) both in my company were killed, the former having his arm cut off....' He added 'My own opinion, formed quietly after this affair, is that had this determined rush taken place a few minutes earlier, there would have been a grave disaster, for the enemy came on with great determination, and appeared to rise from almost under our feet, and were in amongst us as soon as our cavalry were. It was known that the enemy were about, but apparently no notice was taken.

On 3 April the 3,000-strong column set out for Tamaii which they found empty and deserted with the wells filled in. Further probes were made towards Handoub and Otao to cover the laying of the railway tracks which reached out for eighteen miles from Suakin. Then came word that the government had decided that the whole enterprise was once more to be terminated. The railway was left unfinished to rust away, the Marines returned home and the Sudan was abandoned to the Mahdi's bloodbaths for the next eleven years.

This whole period was an exciting and palpitating one for the young men of Victoria's reign. They endured hardships and separations from their families, often for many years, which would be unthinkable these days. But there was a comradeship in their campaigning, coupled with an exhilaration and piquancy for exploring unknown lands that is unlikely ever to be matched in soldiering again. When they did return from overseas they were greeted like heroes, but their lives at home seemed dull and spiritless ever afterwards.

The Royal Marines defending at Tofrek on 23 March 1885. Along with the Berkshire Regiment and under the command of Sir John McNeil, they were manning the 'zareba' when they were attacked by hordes of Arabs, which they drove off with remarkably few casualties. Although the Arab attack was well planned they were confused by the defence in depth concept.

CHAPTER 6

1885-1913
PREPARATION AT SEA

'Their record is second to none. I have been with them on Active Service, on Police Service, in Daily Routine and in Gales of Wind - I had them with me everywhere, and I tell you there is nothing like the Royal Marines.'

Admiral Lord Charles Beresford.
30 June 1909.

Lower Burma was annexed in 1853 after the second Burma War but the hinterland had been largely left to its own devices. However, the accession of King Thibaw saw a renewal in the growth of activity against British trade. A continuing deterioration of the situation finally led to the despatch of a column in November 1885. The expedition, commanded by General Prendergast VC, embarked for Rangoon on the 6th of that month. Along with three infantry brigades, six artillery batteries and some sappers, there was the usual naval contingent with its complement of Royal Marines from ships of the station.

The whole force was conveyed in a flotilla of river steamers and 'flats' up the Irrawaddy from Rangoon to Prome, where they fought their first engagement on 16 November when a series of batteries were stormed and quickly overrun. There followed a number of victorious encounters and landings, each block being taken along with a number of prisoners. Soon the King was asking for peace as the only way to halt the British advance.

But he had left it too late and the force pushed on rapidly until, to the surprise of everyone who knew what hardy fighters the Burmese were, Mandalay fell on the 28th and King Thibaw capitulated. On New Year's Day 1886 this lightning war was over and the annexation of the whole of Burma to the Empire was announced. Compared with the frustrating end suffered by General Wolseley on the Nile in the same year, this second 'River War' was a triumph, yet it is

A scene at Flathouse Quay Portsmouth in 1896 showing, on the left, the training ship hulks of HMS Duke of Wellington *and* Marlborough. *On the right are ships in commission including HMS* Asia. *On the quay in the foreground are RM Light Infantrymen wearing white duck tunics and blue serge trousers, while in the background are stoker training classes.*

The old Crinoline Church at Eastney c1890, now the site of the officers gardens. It was originally built as a hospital for the Crimea and on its return was bought by the Admiralty and erected at Eastney. It held 800 people and was replaced in 1905 by the present Church of St Andrew's.

largely forgotten while Sir Garnet's achievements were trumpeted to the world. Although the regular Burmese army had been defeated, Burma was not wholly subdued. This led to a whole series of campaigns the outcome of which was not fully resolved for another four years.

The Gambia had been settled by the British as long ago as 1588 but in 1843 it was separated from Sierra Leone, included in the West African Settlements in 1866, and again made a separate colony in 1888. It is merely a finger of land fringing the Gambier River and surrounded by French Senegal. There was frequent trouble from the tribes at the head of the river around Tabadian, and Royal Marines formed part of a punitive expedition mounted in 1891.

In 1893, sixty-eight Royal Marines out of a complement of 98 were drowned when the battleship *Victoria* was in collision in the Mediterranean. The *Camperdown* rammed the flagship during complicated manoeuvres off Beirut on 22 June and she sank in ten minutes. As a result of this tragedy, swimming became a compulsory part of Royal Marines training, and swimming baths were built at Deal and Eastney.

In the Gambia during 1894, Lt-Col A D Corbet RMLI had just 50 Royal Marines and 50 native troops of the West India Regiment with which to defend their entrenched positions at Subaji. They were attacked by an estimated 1,500 Mandigo tribesmen who crossed the Gambier River, but steady fire broke up the native charges and they retreated back to their own territory. Corbet followed up this smart little encounter and promptly burnt down the main Mandigo village of Bunjur.

The Marines were present on the West African coast once more on the Brass river in 1895, when the coastal crown colony of Sierra Leone was expanded by the establishment of a protectorate over the hinterland around the area of Magburaka, Kamalu and Kabala in 1896. They also landed for a second time at Zanzibar that year.

During an uprising on the island of Crete in 1897, appeals for help were met by the Mediterranean Fleet, who landed a detachment of their marines under the command of Major J H Bor. They interposed themselves between the feuding factions until peace could be restored.

The Kingdom of Benin lay deep in the jungle of what is now Nigeria, and was largely undisturbed. Unfortunately a visiting party filled with missionary zeal determined on spreading the 'light' even to that remote corner of the Dark Continent. Their motives were misunderstood by the tribesmen, and seen only too clearly by the witch doctors who foresaw the end of their influence. They therefore used their charms, spells and potions to whip the natives into a warlike frenzy and the missionaries were slaughtered. To avenge such an unprovoked assault on a peaceful party, a chastising expedition was mounted under the command of Rear-Admiral Rawson. He naturally in-

The Officers Mess Room at the Royal Marines Barracks Stonehouse in 1896 set for a formal dinner after the presentation of new Colours. Other Colours hanging in the Mess Room are those of 1827 and 1858. The Mess was badly bombed on the night of 11/12 March 1942 when three people nearby were killed.

cluded a force of marines to head it, 100 from the RMLI led by Capt J Byrne and an RMA detachment of 200 gunners commanded by Lt F B Diblee. The column proceeded up the river towards the capital, successfully fighting off two full-scale attacks at Oglobo and Awoko on the way. They reached Benin City on 18 February and found the King and his followers had abandoned it and fled into the bush to escape punishment. All that remained were deserted huts full of beautiful native artefacts from an earlier civilisation, and many of these were brought back home when the city was put to the torch.

The final settlement with the wild Dervish tribes of the Sudan was undertaken in 1896-97, after years of turmoil had led the Khalifa (the Mahdi's successor) to all manner of excesses and the butchery of many of his followers in wars with the Abyssinians. The reconquest of the Sudan was largely carried out to

HM Queen Victoria presents a new stand of Colours to the Portsmouth Division RMLI on 22 August 1894 at Osborne House, her home on the Isle of Wight. At the ceremony Her Majesty said 'They carry on them the Badge of my uncle, King George IV and the motto defining your services by land and sea. I am confident they will always be safe and honoured in your keeping'.

New Colours were presented to the Plymouth Division RMLI on Brickfields by the Honorary Colonel of the Corps, HRH The Duke of Edinburgh, on 3 July 1896. They were received by the Colour Officers, Lts R Prynne and H M Howard, in front of the whole battalion which was drawn up on parade.

Portsmouth Division RMLI on parade at Forton Barracks Gosport in October 1882. This was the year of General Wolseley's Egyptian Campaign in which a large number of marines took part.

Afloat in the 1890s, there was little opportunity to exercise. Royal Marines were allowed to use the quarterdeck for 30 minutes a day even in dry dock in Portsmouth, and formal bayonet exercises were part of their curriculum.

forestall the infiltration by other powers, particularly France and Italy, and to maintain the British presence through the outward façade of the re-establishment of British rule. The Egyptian army had been completely re-trained, re-armed and reorganised with British officers, modern magazine-filled rifles, and a railway to support it as it progressed up the Nile. The whole campaign went like clockwork. The only Royal Marines involved were Capt H Slessor RMA, who commanded the 1st Egyptians, Capt G E Matthews RMLI the 14th Soudanese and Lt-Col C V

NCOs, Gunners and Buglers of the Royal Marine Artillery in various orders of dress worn overseas in 1895. Of particular interest are the different forms of headdress. Three wear white helmets with pugarees, four have the pre-1905 white tropical helmet with Bursting Grenade badge, ball top and chin chain, whilst the remainder wear the pill box hat, two with white covers.

F Townsend RMLI the 12th Soudanese. Col Townsend served with much distinction with the army, including service with the Marine company of the Guards Camel Corps in 1884, commanding the garrison at Chitral in 1895, and finally, as a major-general, becoming the famous defender of Kut in Iraq in 1915. There was also a small RMA detachment of a sergeant major, nine corporals and bombardiers under Capt H Oldfield who had been sent out from England in June 1896 to train and supervise the Egyptian gunners aboard the fleet of river gunboats which accompanied the expedition. One of these CSgt Jenvey RMA had the last word on all the various campaigns in the Sudan, when he placed the 300lbs of gun-cotton which blew asunder the tomb of the Mahdiat Omdurman.

In a manner similar to that in which the American colonies had reacted once a British force of arms had removed an enemy threat from their homelands, so the Dutch descendants of the Cape's South African hinterlands turned against the British after the Zulu threat had been eliminated. Thus the second Boer War of 1899-1902 inevitably followed the first. A contributory factor had been the lack of a decisive Boer defeat in the earlier war. Although it might have been possible to inflict such a defeat on the Boers, (indeed many, including the Queen, had thought it desirable) it had not been administered. Therefore the Boers felt confident that they could repeat the

A Bill poster for an RMA concert on 29 April 1892. At this time the theatre was part of the canteen, but a new theatre the Globe Theatre was built from canteen funds in 1899.

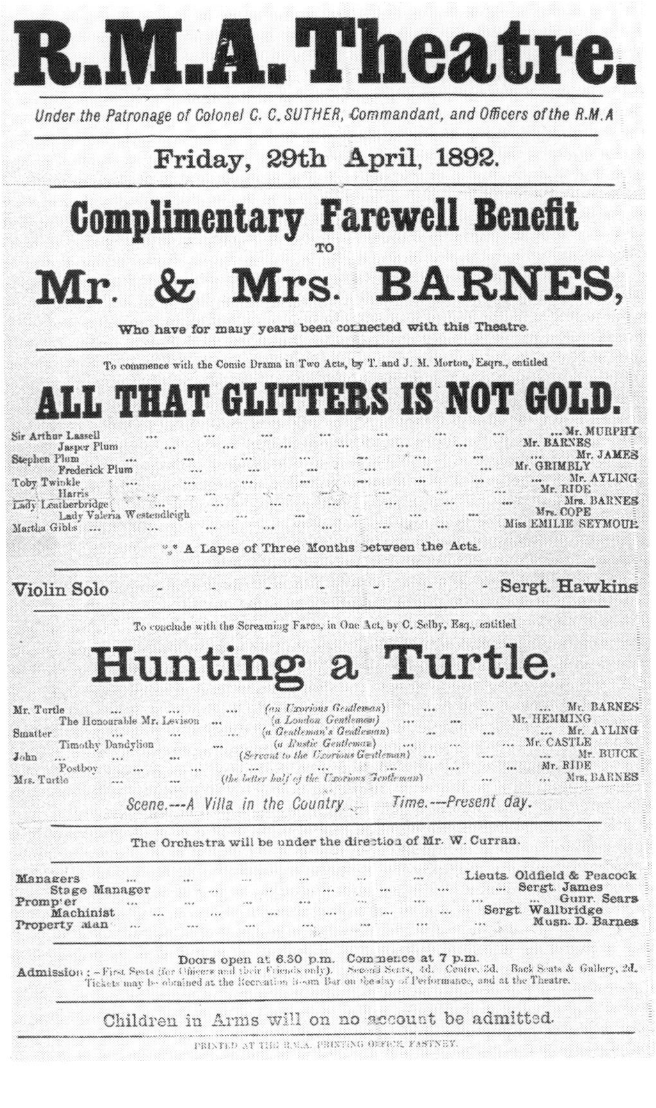

R.M.A. Theatre.

Under the Patronage of Colonel C. C. SUTHER, Commandant, and Officers of the R.M.A

Friday, 29th April, 1892.

Complimentary Farewell Benefit

TO

Mr. & Mrs. BARNES,

Who have for many years been connected with this Theatre.

To commence with the Comic Drama in Two Acts, by T. and J. M. Morton, Esqrs., entitled

ALL THAT GLITTERS IS NOT GOLD.

Sir Arthur Lassell	Mr. MURPHY
Jasper Plum	Mr. BARNES
Stephen Plum	Mr. JAMES
Frederick Plum	Mr. GRIMBLY
Toby Twinkle	Mr. AYLING
Harris	Mr. RIDE
Lady Leatherbridge	Mrs. BARNES
Lady Valeria Westendleigh	Mrs. COPE
Martha Gibbs	Miss EMILIE SEYMOUR

*** A Lapse of Three Months between the Acts.

Violin Solo - - - - - - - Sergt. Hawkins

To conclude with the Screaming Farce, in One Act, by C. Selby, Esq., entitled

Hunting a Turtle.

Mr. Turtle	(an Uxorious Gentleman)	Mr. BARNES
The Honourable Mr. Levison	(a London Gentleman)	Mr. HEMMING
Smatter	(a Gentleman's Gentleman)	Mr. AYLING
Timothy Dandylion	(a Rustic Gentleman)	Mr. CASTLE
John	(Servant to the Uxorious Gentleman)	Mr. BUICK
Postboy		Mr. RIDE
Mrs. Turtle	(the better half of the Uxorious Gentleman)	Mrs. BARNES

Scene.---A Villa in the Country. Time.---Present day.

The Orchestra will be under the direction of Mr. W. Curran.

Managers	Lieuts. Oldfield & Peacock
Stage Manager	Sergt. James
Prompter	Gunr. Sears
Machinist	Sergt. Wallbridge
Property Man	Musn. D. Barnes

Doors open at 6.30 p.m. Commence at 7 p.m.

Admission: – First Seats (for Officers and their Friends only). Second Seats, 4d. Centre, 3d. Back Seats & Gallery, 2d.
Tickets may be obtained at the Recreation Room Bar on the day of Performance, and at the Theatre.

Children in Arms will on no account be admitted.

PRINTED AT THE R.M.A. PRINTING OFFICE, EASTNEY.

The RMLI detachment of HMS Immortalite *in marching order prior to a boarding exercise in 1897.*

Crossing the Line ceremony aboard HMS Terrible *on the way to the Far East, via South Africa in 1899. Those officiating at the ceremony, which has its origins in pagan religious rites, had always previously crossed the equator. The Bears, who were responsible for catching the 'novices' and ducking them, were often provided by the Royal Marines detachment, their hairy coats sometimes exuding a strong smell of Stockholm tar.*

'Col Crease turning the defences'. Col J F 'Toby' Crease was a renowned character who joined the Corps in 1854. Amongst other achievements, he designed a system of drainage for Portsmouth. In 1891 he was of the opinion that the defences of Portsmouth could be taken in reverse and obtained permission from the local G-O-C, The Duke of Connaught, to attempt to do so with Royal Marines. On Easter Monday he landed with a Battalion each of RMA and RMLI just south of Hayling Island ferry and they marched across the mud with flat boards on their feet to capture Fort Cumberland which was being held by Militia Artillery. They then marched through Milton to Hilsea, capturing the Colonel RA and the CO of the Hants Volunteer Artillery en route as they were riding home for lunch.

trick against the 'soft' British and so hostilities were renewed in October 1899. It immediately became clear that the jealousy of other European states, notably Holland and Germany, would ensure that the Boer's demands were enlarged out of all proportion. A general hate campaign against the British was mounted and this, in turn, ensured that the Empire looked to its 'moat' and mobilised the Royal Navy (as strong as the combined fleets of her two closest rivals) as a precautionary measure in case pompous rhetoric was replaced with war.

The decision meant that the Royal Marines remained chained to the fleet, but they did get one chance to fight ashore. Like General Colley before them, and many other regiments and hitherto invincible generals afterwards, the Royal Marines had to re-learn that fighting a few skilful, well-armed marksmen, who preferred hit and run tactics to an orthodox stand-up fight, was a very different proposition from taking on hordes of Zulu, Dervish or Ashanti warriors.

This salutory lesson was first imparted to the Royal Marines on 25 November 1899. Two RMLI companies and one RMA company, with a total of 190 officers and men, had been formed at Simonstown in South Africa from the complements of the *Doris, Powerful and Terrible*. Their task was to accompany the Naval Brigade's sailors and four 12-pounder guns, which were sent by rail to reinforce Lord Methuen's relief column on its way to raising the siege of Kimberley. They had acted in support at the Battle of Belmont on the 23rd and two days later spearheaded the assault on Boer positions dug in atop Graspan kopje. As usual, the British advanced across the vast featureless veldt in open order, completely exposed to the accurate long-range fire of their concealed enemies. Bunker Hill in the past and the Somme in the future showed that many British generals knew no other way!

The Royal Marines companies assembled at 0700hrs and moved forward from their start line

A Team of RMA about to receive the First Prize for Bayonet exercises at the Naval and Military Tournament in Cape Town in January 1902. The principal judge on the left is a Foot Guards Regimental Sergeant Major.

NCOs and Men of the Royal Marine Artillery are presented with South African War Medals by Colonel Commandant Guise Tucker in front of a parade of over 1,600 Marines at Eastney on 8 March 1902. On the same day HM King Edward VII presented China Medals to 7 Officers and 43 NCOs and Men on parade at the Royal Naval Barracks, Plymouth.

A Royal Marines Light Infantry guard of honour for the visit of the Duke and Duchess of Connaught to Simonstown in 1910. A permanent Royal Marine garrison was maintained in South Africa in the early art of the twentieth century.

Right:
A Royal Marines Legation Guard was maintained in Peking in the 1890s, along with Guards from other nations. The small area of the British Legation to be defended was about 1,000 yards square. The garrison knew they would always have to fight to a finish in the event of attack by any barbaric hordes. This 1894 photograph shows the RMLI Guard.

towards the enemy-held crests nearly half a mile away as soon as the preliminary bombardment had stopped. Each marine was only four paces from his neighbour, and thus bunched they made a series of rushes. The country across which they advanced was plain grassland, about 18" high, with just a few scattered ant-hills to break the ground. Not a shrub nor a hillock offered any protection. The skyline, until then seemingly abandoned and empty, suddenly came alive with Boer riflemen. A heavy and accurate fire swept through the advancing British line, marines falling at every step, including the commanding officer, Major J H Plumbe RMLI, Capt G Senior RMA and ninety others, of whom five were killed outright.

Lt (later the Rt Hon Sir) Leslie O Wilson RMLI later recalled bitterly how, having braved all this and having suffered almost 50% casualties and still pushed on, the marines were denied the fruits of their valour:

> When we reached the foot of the kopje we were in comparative safety, out of sight of the enemy. Our men rushed the position with fixed bayonets, but by the time they reached the top of the kopje, all the Boers had fled in carts or on horseback carrying their dead and wounded with them.

Although a second Royal Marine battalion was to form part of a further naval brigade during the Natal campaign, they were not destined to play a similar leading role in any of the subsequent battles in a way that became more and more a guerilla-type conflict.

Ever since the end of the second China War the Celestial Empire kept to its treaties, although their crushing defeat at the hands of the Japanese had shown that their expertise still did not match their hardware. At Bejing (Peking) all the foreign nations maintained Legations in the European quarter compound which formed an inner walled annexe of its own, just outside the sacred Imperial City with its own massive defences, but within the Tartar walls which were surrounded by suburbs.

Just like the *Ronins* in Japan a quarter of a century earlier, a nobleman, the Prefect of Tsau Chaou, had raised and financed an extreme nationalistic group, known as the *Patriotic Harmony Fists* or *Boxers*. Although opposed to the rottenness of their own Imperial Dynasty as well as the christian Chinese, they also railed against all 'foreign devils' as having an evil and corrupting influence. Like the Mahdi they wanted to rid their homeland of all such cancers by force, given half a chance. This movement grew and murders and disturbances against European and American traders and factories became widespread.

In order to provide the diplomats with some protection it had long been the custom for detachments of troops to be stationed in the Legation compound, which enjoyed many extra-territorial rights. In view of the deterioration of the situation, it was decided to send the 'Winter Guard' some time ahead of schedule, and thus on 31 May 1900, the Americans, French, Austrians, Russians, Germans, Italians, Japanese and British, plus many civilians, reached the compound. The British detachment comprised three officers and 79 men of the RMLI and, as they swung through the gates of the city, they had to pass through streets lined with thousands of resentful Chinese Boxer sympathisers. There was not much doubt that they were walking into a trap, and their Chinese guide warned them that they would not leave the same way as they had entered, but 'Up Top'. They arrived none too soon, for less than a month later the Boxers reached the city and soon began

RMLI NCOs of the Legation Guard, Peking in 1900 during the Boxer Rebellion. They are (left to right) Cpl Gowney, Cpl Shepherd, Sgt Saunders, Sgt Murphy, Cpl Gregory, Sgt Preston (who won the Conspicuous Gallantry Medal) and Cpl Johnson. Seated is Signalman Swannell RN.

Royal Marines, a British seaman and two US Marines manning a Maxim machine-gun outside the front gate of the British Legation in Peking, nicknamed 'Fort Halliday' after Capt L S T Halliday RMLI, who won the Victoria Cross there on 24 June 1900. 'Our Betsy', the International Gun, was so called because the barrel was British, it was mounted on an Italian gun carriage, fired Russian ammunition and was manned by an American gunner.

enlisting large sections of the populace, killing many Chinese Christians and burning and looting their homes and shops.

The Imperial Government, as wily as ever, hoped to gain time by turning the wrath of the Boxers against the foreigners, of whom they had little love anyway, and away from themselves. On 19 June they ordered all foreigners out. After a meeting of all the senior diplomats, the German minister could wait no longer and set off with his interpreter by sedan chair to negotiate, but his Chinese escort turned on him and murdered him en route. Plans had already been made to co-ordinate the defences of the Legations and barricades were built, makeshift cannon mounted and patrols organised by all the nations shut up in the city. There was ample water but little food.

For the next 55 days the desperate defenders held out against the continual encroachments of much of the Chinese population. The Imperial troops threw off any disguise at neutrality and openly supported the Boxers. Houses which hemmed in the Legation compound were set ablaze and a steady infiltration, coupled with probing attacks, continued day and night. Many gallant sorties were made against the Chinese besiegers and here the Royal Marines, on

Marines from HMS Orlando *march though Tsentin which was beseiged by the Boxers during the summer of 1900. They first landed on 28 May 1900 and again on 31 July. With four Companies under Major E V Luke they set off for Peking on 3 August. The ancient Chinese city which stands 30 miles up the Pei-Ho river was under constant bombardment and the Royal Marines Light Infantrymen, along with a multi-national force withstood continual bombardment and attacks.*

The Mess Room of the old Sergeants Mess at Eastney c1906, which was situated on the north side of the Drill Shed. This was vacated in 1967 when the mess moved to its present site on 7 September.

Barrackrooms in the early part of the 20th century were occupied by between 16-20 Marines. In Forton Barracks in 1901 this room was decorated with photographs and knicknacks. The tables and benches were scrubbed daily and the fireguard and dixies burnished, a practice which continued into the Second World War. There were no lockers and kit was packed or stacked with the bedding rolls and on shelves above the bed.

many occasions, fought side by side with troops of other nations, including the United States Marines Corps, who had three officers and 53 men present.

The senior German officer of the Legation guards who had assumed overall command panicked during a Chinese attack whereupon the British minister in Peking, Sir Claude MacDonald, took over and co-ordinated the defence in which much heroism was displayed. It was here that Capt Lewis Stratford Tollemache Halliday RMLI earned the Corps another Victoria Cross on 24 June. A heavy attack had been made by a combined force of Boxer and Imperial troops on the east wall of the British Legation. They set fire to the West Gate of the south stable quarters infiltrating into the buildings in the lee of the wall. In order to allow the defenders to extinguish the fire, a sortie was made by Capt Halliday and twenty men.

They knocked a hole in the Legation wall through which the marines, led by Halliday, stormed the recently-occupied buildings. He was shot through the shoulder as he burst his way into the first house. Despite this wound Halliday used his service revolver to good effect, shooting dead four out of the five Boxers in quick succession, ordering his men to deal with the rest. This they did with a will after his men had urged him to go to the Legation hospital for treatment. He refused to let anyone accompany him so as not to deplete their numbers. Pte Sawyer was seriously wounded, Pte Layton had bullet holes through his serge tunic and trousers without receiving a scratch, while Cpl Gregory had his bayonet shattered by a bullet. But the marines achieved their aim, driving the enemy from their position.

There were many brave deeds performed, one of which was when Sgt J E Preston RMLI won the Conspicuous Gallantry Medal when he captured a Boxer flag during a sortie to clear part of the Legation wall on 14 July, while the next day Capt B M Strouts RMLI, who had been in command of the Royal Marines detachment, was killed at the head of another storming party.

Down on the coast the International fleet landed a large force of 2,000 men at the mouth of the Pei-Ho river to relieve the Legations, and these left on their rescue mission by rail on 9 June. Included in this column, under the overall command of Vice-Admiral Sir Edward Seymour, were a further 213 marines. When they started out their briefing had been unclear and they had little idea that the allegiance of the Imperial troops had changed, but when they were 25 miles from the city they discovered the railway lines had been torn up both ahead of them and in their rear, and an overwhelming force awaited them. They had little choice but to fight their way back to the main base at Tientsin on foot. It was night when this little column marched past the Hsiku arsenal some ten miles from the town and on the opposite bank of the river. Here the Chinese sprang a treacherous trap. They hailed the force and, on being told its true identity, replied that all was well. In an instant, as soon as the international force was exposed, they opened fire with every weapon they had. For a few moments there was confusion and it seemed that the

Sportsmen of the Day - 1900. Royal Marine Artillery champions outside Eastney Gymnasium. From left: Cpl W McFadden (Army & Navy Heavyweight Boxing Champion), Sgt S Gadcombe (Heavyweight Lifting) and Cpl G R Waterman (5-mile Bicycle Race Winner).

The old Eastney Main Gate with its three pillars c1904. Owing to the heavy transport passing through these gates in the 1960s, the centre pillar was removed and some of the bricks were sold in aid of the Museum Fund. The sign on the pub opposite the gate still shows the three columns.

A group of Marines pose in front of the wooden hutted mess at the Taku Forts in 1901. These forts were powerful fortifications at the mouth of the Pei-Ho River which flowed down from Peking. The Royal Marines first took them in 1858 and subsequently found them to be a severe obstacle to the relief column on its way to the Peking Legations in 1900.

Royal Marines Light Infantry line the streets of Cowes, Isle of Wight, for the funeral of Queen Victoria on 2 February 1901. Also lining the route from Osborne House to Trinity Pier, East Cowes, were four Companies of Royal Marine Artillery with the Massed Bands under 2nd Lt George Miller RMLI, who had been the first Bandmaster to be commissioned.

whole column would be destroyed, but Major W N G Johnstone RMLI led his marine detachment back the way they had come and across the river to the rear of the arsenal. The marines then carried out a bayonet charge, routing the Boxer garrison and the whole column took refuge there themselves. If they were to be besieged there, then there was little hope for the small garrison at the base camp at Tientsin, where Major E V Luke RMLI and Lt Armstrong RMLI had been left with some sailors and a few German, Russian and French troops. They, too, were cut off and withstood numerous assaults.

With three groups now surrounded along the Pei-Ho river it was clear that a proper relief force, strong enough to deal with the Boxer rebels would have to be sent up river. First, of course, they would have to take the Taku Forts, for long a thorn in the marines' side. It was the same old problem, as the intervening 40 years had seen the modernisation of the forts and the installation of Krupps artillery pieces. In addition four German-built destroyers lay anchored below each of them, armed with quick-firing guns. Accordingly a landing force was sent ashore quickly on 17 June to forestall any Chinese moves and they

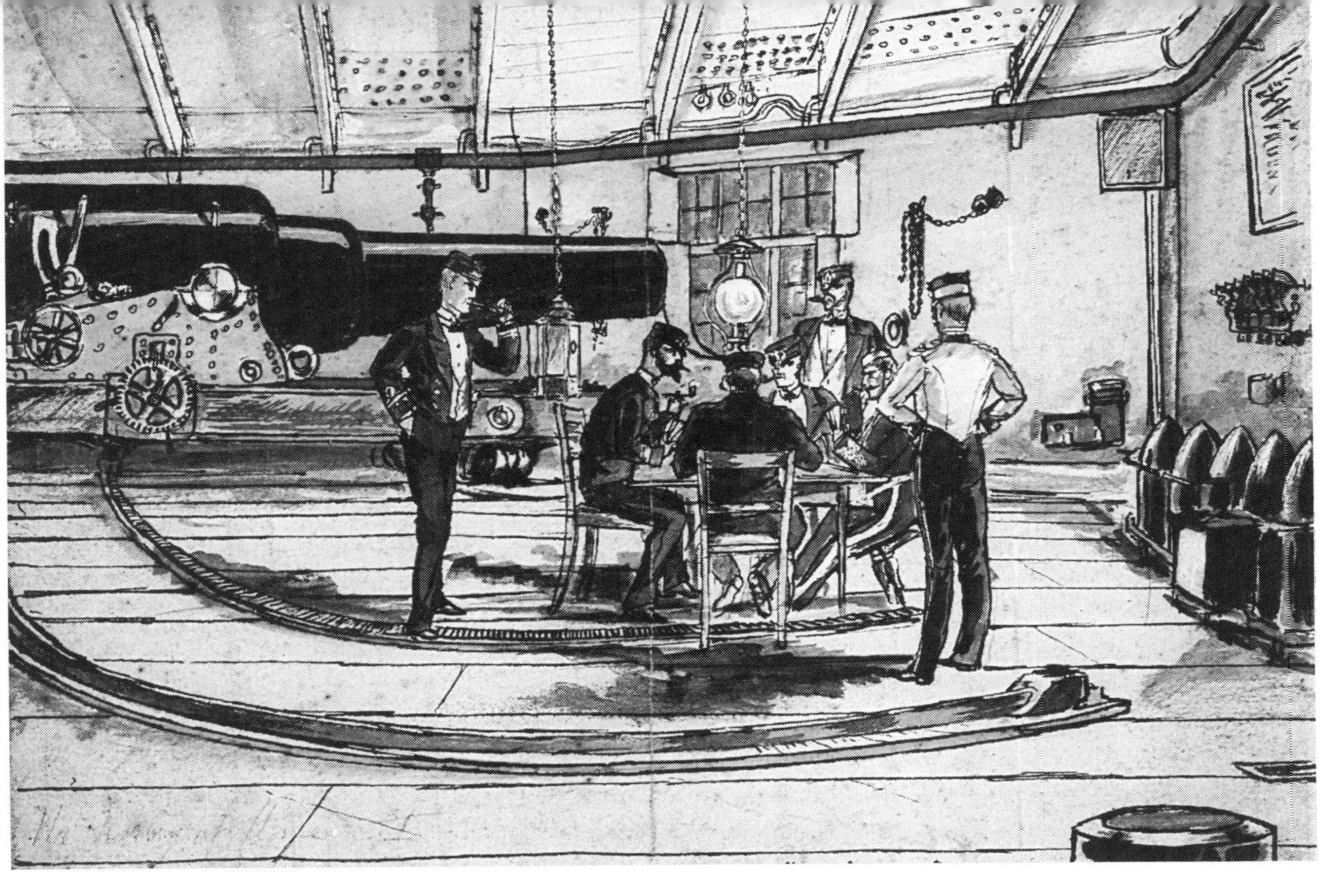

captured the forts and took over the warships. By 4 August, 20,000 men had been assembled and they slowly fought their way through every little enclave in turn; Pei Tsang on the 5th; Yang Tsun on the 6th; Tungchow on the12th; and they eventually reached the walls of Peking on 14th August to raise the siege. Of the 13 officers and 545 marines who participated in the Boxer uprising, 4 officers and 36 men were killed with 2 officers and 79 men wounded.

In a way it marked the end of a unique period, for although it was not the last of the little wars of the Colonial age it was almost so, and fittingly apt to put a seal on that golden era. Queen Victoria was to pass away peacefully at Osborne House on 22 January 1901 and the Royal Marines bands that played at her funeral procession on the Isle of Wight might well have been playing for the funerals of a whole generation, for just around the corner lay the bitter disillusionment of the Great War.

In Edwardian days, as was the case up to the 1950s, one third of the armament of all the fleet's big-gunned ships, battleships, battle-cruisers, armoured and light cruisers was manned by Royal Marines. Both 'Red' and 'Blue' marines served afloat more and more as their land assignments waned. The average strength of the Corps at the beginning of the 20th century was 15,000 officers and men, and the three Grand Divisions of Chatham, Portsmouth and Plymouth remained. The Royal Marines Depot was at Walmer in Kent and the Royal Naval School of Music was established at Eastney barracks in 1903. This co-ordinated all sea-going bands, which had previously been mostly civilian, many of them Maltese. The Divisional Bands remained permanently in their barracks.

'A Quiet Rubber Afloat'. A cartoon depicting officers playing bridge on the quarterdeck of the battleship HMS Nelson *whilst visiting Suva, Fiji in 1888.*

From the unpublished drawings of Col C E Field

A typical battleship of the day (pre-Dreadnoughts as they were known after the arrival of that all big-gun ship rendered all her contemporaries obsolete in 1906) might be taken as the *Canopus*. She mounted the standard four 12-inch, twelve 6-inch armament of the day, having a total complement of 737 officers and men of whom 50 would be RMLI and 53 RMA. These would be something like 1 captain, 2 lieutenants, 1 colour-sergeant (with the courtesy title of sergeant- major), 2 sergeants and about 100 marines. Their basic duties, apart from manning the armament and providing guards of honour, were to act as sentries in various parts of the ship and provide wardroom attendants. There were also specially-created rates for marines such as barber, butcher and lamp-trimmer, which first appeared in 1886 and paid an extra tuppence a day, but they did no boat work. The sergeant-major was responsible for reporting to the captain each day that all clocks were wound. Otherwise they would share in the general drills and evolutions with the sailors and perform an hour's arms drill and bayonet practice on the quarter-deck. When the unavoidable chore of coaling ship was piped, the marines usually handled the sack barrows

The first 34 personnel of the Royal Naval School of Music when it was formed on 22 July 1903. Chief Bandmaster (later Major) H E Lidiard, who also held the appointment of Inspecting Bandmaster of the Royal Navy, is in the centre. He had already spent 26 years in the Royal Navy coming to Eastney with his band from HMS Impregnable. *Known as the 'Father of the Band Service', he retired in 1919 and the housing estate which is now on the site of the old RNSM is named* Lidiard Crescent *in his honour.*

The Plymouth section of the 'Woolwich Folly'. This extravagant silver centrepiece was cast in silver and electro-plate at a cost of £450 in 1854 by the officers of the Woolwich Division. On the Division closing down in 1869 the 'trunk', the largest section, was transferred to Plymouth Division while the two 'branch' sections went to Chatham and Portsmouth. Its nickname is said to derive from its enormous weight which broke the mess-room table, but it is probable that the name refers only to the expense.

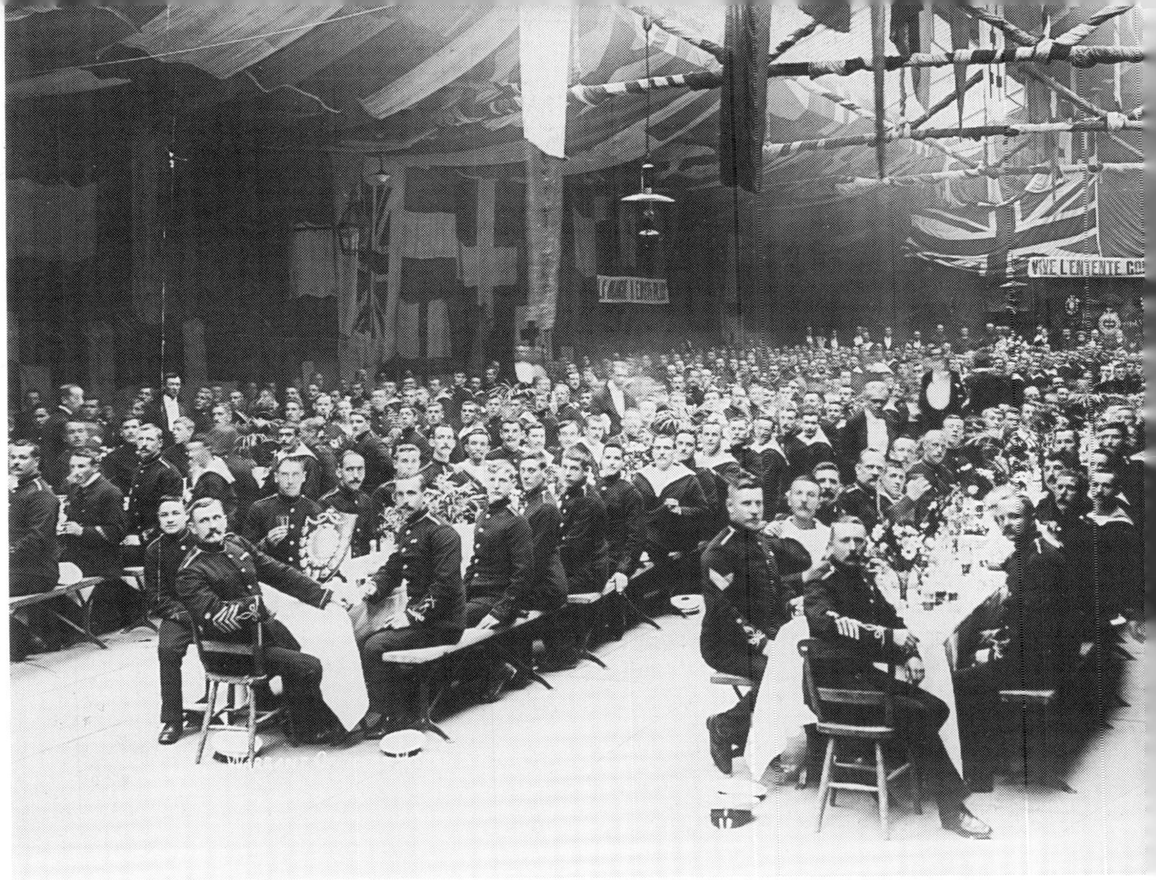

The Band of the Royal Marine Artillery lead the parade along the 'Golden Mile' to the Crinoline Church at Eastney in 1903. This band was separate from the Royal Naval School of Music, established at Eastney that year, and in 1904 the Band was honoured when HM King Edward VII decreed that it should become the permanent Royal Yacht Band.

Top right:
A decorated Eastney Drill Shed was the setting for a reception for sailors of the French Fleet given by the Warrant Officers, Senior NCOs, Corporals and Gunners on 12 August 1905. The Drum Major and the Bugle Major are at the ends of the tables.

The Royal Marines Band play on the quarterdeck of HMS Hercules *after Sunday morning service while the ship was in Weymouth in 1912. Full Dress was worn and both an RMA officer (Blue Marine) and an RMLI Officer (Red Marine) can be seen in the foreground.*

which the rest of the crew loaded from the colliers alongside.

Their quarters, known as the 'barracks', were by long tradition always between the officers and the crews quarters in order that a neutral and disciplined force should be in a strategic position in case of mutiny and unrest. However, as often as not, the marines and sometimes their officers, sympathised with the seamen's grievances and were usually too few to prevent mass disobedience. Past (and present) experience in such matters made it plain that prevention of such unrest depended more on the way the officers handled their men than on the bayonets of the marines.

The young officer candidates for the Corps were selected from the ages of 16 to 18 for the RMA and from 17 to 19 for the RMLI. If a university graduate wished to apply for an RMLI commission then the upper limit was extended to 23. The entrance and other examinations were in line with those in the

The Royal Marines Memorial in the Mall, near Admiralty Arch, which was 'Erected by the officers and men of the Royal Marines in memory of their comrades who were killed in action or died of wounds or disease in South Africa and China 1899-1900.' It was sculptured by Capt Adrian Jones, a retired Army Veterinary Corps officer, and was unveiled by HRH The Prince of Wales (Colonel in Chief) on 25 April 1903. Originally it stood in St James's Park, but was moved in December 1940 to make way for the Citadel, stored at Battersea until 1948 when it was re-erected in its present position. A short service and parade is held there each April.

army, with the RMA being the same as those sat at Woolwich and the RMLI those taken at Sandhurst. If successful the young candidates were assigned to the Royal Naval College at Greenwich with their naval equivalents.

On leaving Greenwich RMLI officers joined a Divisional HQ to complete their military training, while the 'Blue' marines officers had to attend courses at the Royal Navy's gunnery and torpedo schools. Under

Col A F Gatliff presents Long Service and Good Conduct Medals to Royal Marine Light Infantrymen at Forton Barracks in 1909. The men are wearing 1902 pattern tunics and 1905 pattern white helmets.

Admiral Fisher, these became the 'stone' frigates of HMS *Excellent* on Whale Island and HMS *Vernon* at Portsmouth. Officers were paid the handsome sum of £69 per annum as a second lieutenant, and if they reached the dizzy heights of a colonel commandant they received £730 a year. As they were probably only employed for six to twelve hours a week on Corps duties, this was not hard-earned money but it was stultifying because ambitious officers could reach no higher rank those days.

Under Lord Fisher's sweeping reforms new colleges were built for the navy at Osborne and Dartmouth. Two years were spent at each followed by six months in a training cruiser. Fisher, who had a blind spot as far as the Royal Marines were concerned, merely wanted them to boost the number of sea-going officers without regard to their own special standing and traditions. He tried to force reforms on them including making Royal Marines officers follow the same initial routine as naval officers, the common entry or 'Selbourne Scheme' being introduced in 1907. On reaching lieutenant's rank, individuals could choose the marines if they so wished, thus beginning their specialisation. This meant that any officer who chose the Corps would forego the chance of commanding a ship or squadron. This was one of Fisher's ideas that back-fired. Few young officers emerged from the preliminary training to choose what they considered to be the dead-end job of a marine officer. At an age when many were thinking of marriage, these young men saw no merit in tying up their future.

Although Fisher declared his scheme unstoppable by 'prejudice, Parliament, Satan or even, beyond all these, the Treasury itself', the Royal Marines stopped it within three years. They had already seen their military role declining for they were no longer sent on military manoeuvres with the army as they had at Aldershot between 1892 and 1899, and regarded the scheme as the last straw. They wished to be first-class sea-going soldiers, but they were being turned, they felt, into second-rate naval officers. Their representations were justified, but more telling was the fact that only two officers actually joined the marines while the scheme was in operation. Fisher reluctantly gave way, and from 1912 a separate entry for Royal Marine officers was re-introduced and so it remains, with variations, to this day.

The Corps may have been helped in its fight by Royal Patronage. On 9 December 1882, Queen Victoria's youngest son. HRH the Duke of Edinburgh (later the Duke of Saxe-Coburg and Gotha) was appointed Honorary Colonel. Disappointment had been expressed that he was not appointed Colonel-in-Chief and this was remedied in the next Royal appointment. On 1 January 1901 the Duke of Cornwall and York had become the Colonel-in-Chief of the Corps, an appointment which he continued to hold when he ascended to the throne as King George V nine years later. He had, as Prince of Wales in 1902, led a four-battalion Royal Marine Brigade in review past King Edward VII at the Coronation review. The Royal Family has maintained its close connections and interests in the Corps right up to the present day.

One unhappy side effect of Fisher's experiment was that the Corps was 58 subalterns short at the outbreak of war in 1914. The rank-and-file marines were of a similar professional band of brothers, so the selection of entrants for such a compact and élite force could afford to be exacting. Only the fittest and most physically perfect specimens were chosen from the many volunteers for the Corps, height and body measurements being very severe, while the discards were offered to the navy or army! Those selected regarded themselves as a 'cut-above' the army in all respects, and over the years since Lord St Vincent's patronage, undoubtedly earned a firm reputation as a reliable and trustworthy regiment. Those admitted therefore had justifiable pride in themselves and the Corps, and the long traditions of the Royal Marines quickly emphasised and enhanced this feeling of 'a Corps apart'. The officers might find promotion frustratingly remote and their duties sometimes irritatingly vague, but for most of the other ranks, who came from the lower social orders of Edwardian England, the Corps gave a man pride not only in himself and the Corps, but also in his company or his ship. Volunteers signed on for a minimum of 12 years only qualifying for pension rights after 21 years. They were paid from £21 per annum as an RMLI private with no special rating, to £73 per annum as an RMA colour-sergeant. Everything for those afloat was found, and as time went on for those in barracks ashore too.

Initial drill at the recruit depot was severe and those who were not rejected during their first few weeks were sent to their Divisions at one of the three manning ports and placed on a general service 'rota', from which they were assigned to ships as commissioned. When the big multi-turreted 'Dreadnought' battleships started to join the fleet as the naval arms

Right:
An inspection of the RMLI Depot Band at Deal outside the Officers Mess in South Barracks.

Below right:
One of the traditional duties of a Royal Marines detachment afloat was to provide the ship's postman. Here a Corporal RMLI in full dress tunic comes aboard with the mail while a SNCO in his blue Kersey working tunic and scarlet sash (worn on duty) stands at the top of the gangway. Both NCOs are wearing the Glengarry cap first introduced in January 1870 with the separate Globe & Laurel and Bugle Horn badge.

'Up Spirits' was sounded at six bells in the forenoon watch each day (1100 hrs) and all sailors and marines over the age of 18 were entitled to a tot of rum (or money in lieu). In the late 18th Century this amounted to half a pint per man per day. In 1881 this was reduced to an eighth of a pint and remained so until the rum ration was abolished on 31 July 1970. Senior Ratings and NCOs drew their rum neat, while the rest of the crew had a 'two water / one rum' mixture. It was the Royal Marine butcher's duty to assist in this daily ceremony. This photograph was taken aboard a battleship in Portsmouth on New Year's Eve 1910, the day after coaling ship.

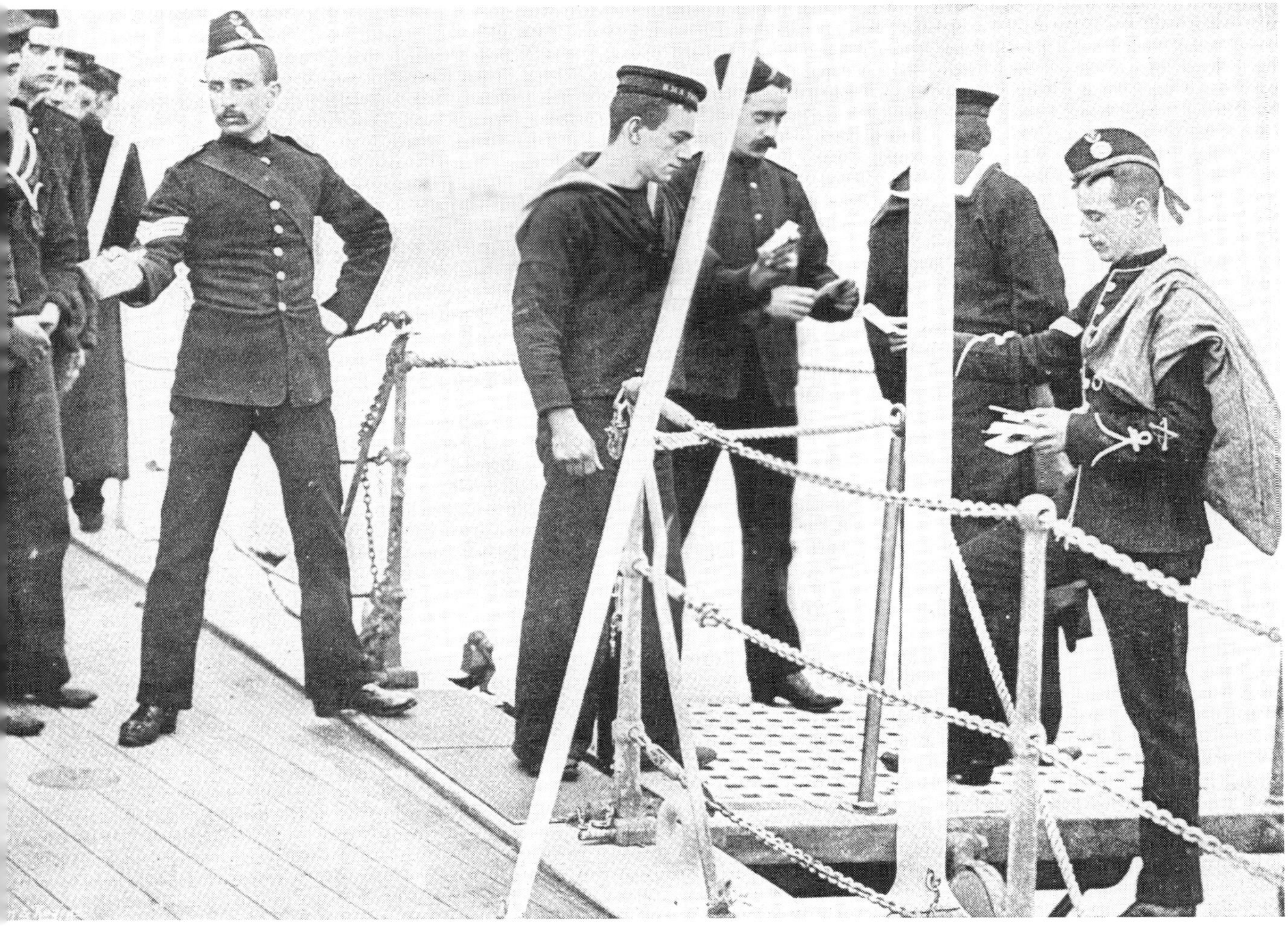

race escalated from 1908 onwards, each battleship had one turret designated to the Royal Marines, usually the after 'X' mounting, the superimposed stern turret. Marine gunners also serviced the secondary armament including the newly-introduced 'quick-firing' weapons designed to beat off torpedo boats and destroyer attacks.

Prominent aboard the larger ships of the fleet after 1903 were the Royal Marine bands, which were always popular both for ceremonial displays and for lighter social functions when parties were given for civilians and visiting dignitaries. The Royal Marine bandmaster had to be a highly versatile musician with a very wide repertoire indeed. The bands were popular with both officers and men because of this, often providing the lowerdeck with about the only social interludes they had in between leave periods. Only once did a bandmaster fall foul of a senior naval officer which led to the much later and infamous *Royal Oak* court martial episode in 1928. This was more of a farce then a major incident, but it certainly made the headlines and brought the whole service into disrepute. During a social function on the quarterdeck of HMS *Royal Oak* in Malta, Rear Admiral Collard, a well known martinet, was heard to swear at the Royal Marine Bandmaster in front of the officers and their ladies, for playing some modern American music. The following day the Admiral was ordered by the C-in-C to haul down his flag, the Captain of HMS *Royal Oak* was court martialled and the Bandmaster asked to leave the service prematurely, but this was not granted. The innocent catalyst of the trouble was Royal Marines Bandmaster Barnacle, a name which the press found irresistible, of course!

Finally, although Fisher wanted to turn the Royal Marines into a kind of ship-borne reserve for the navy, it did not prevent him advocating some crazy combined operations, like the proposed landing of a 3,000-strong force of marines on the island of Borkum, or the storming of the Heligoland fortress

Royal Marines of the Mediterranean Fleet march past King Edward VII and Queen Alexandra when they reviewed the troops in Malta in 1907. The Floriana parade ground on the right was the scene nearly fifty years later, on 29 November 1952, of the presentation of the first Colours to 40, 42 & 45 Commandos by HRH The Duke of Edinburgh.

Top right:
Royal Marines of the Mediterranean Fleet march through Palace Square Valetta, Malta to celebrate the King's Birthday on 3 June 1907. A Royal Marines Battalion of 6 Companies was provided including mounted officers. The Royal Guard (right) was provided by the Rifle Brigade.

Below right:
Royal Marines from the Fleet march through Hong Kong to celebrate the Coronation of King George V in 1911. The Hong Kong Club, which can be seen on the left close to the waterfront is now about 400 yards inland due to reclamation of land.

where the defences were stronger than those at Gibraltar by a third of that number. Even a landing on the firm sands of the north German coast to seize Jade and Elbe forts was considered well within their capability, which in view of the size and power of the German army there, and the range of their modern weapons,would have made Cartagena and Rochefort look like a picnic.

Top right:
The Royal Marine Artillery march past in Review Order on Southsea Common on the occasion of the visit of the Spanish Royal Family in 1905. In the background can be seen the Queen's Hotel, a welcome sight for all naval ships returning from overseas and one of the first recognisable features of Southsea seafront.

Below:
HRH The Prince of Wales, as Colonel-in-Chief, inspects a parade of the Royal Marine Artillery at Eastney on 16 March 1904. After lunching in the Officers Mess, he attended the inauguration of the new Church of St Andrew's when HRH The Princess of Wales laid the foundation stone, under which were placed contemporary documents and coins. The church, which was built of Rowlands Castle red brick with Portland and Bath stone dressing, was dedicated on 17 November 1905 by the Lord Bishop of Winchester.

Below right:
Group of RMA signallers on a three months course at Eastney in 1903. Morse by heliograph and semaphore were widely taught and used, while bicycles were the main mode of transport.

R. ELLIS
C.

Royal Marine Artillery Fleet Reservists at 12-pdr gun drill in the Sea Service battery at Eastney in 1904. This subsequently became Melville Camp and housed the Drafting, Pay and Records Office from August 1960 to March 1973 after the closure of Melville Barracks, Chatham.

Right:
The German Kaiser inspecting a Royal Marine Artillery Guard of Honour commanded by Capt A G Troup at South Railway Jetty, Portsmouth on 11 November 1907, accompanied by the Prince of Wales (later King George V).

Below left:
Coaling ship was the least popular of all naval evolutions. The whole of the ship's company took part in this, including the officers, and it was the one occasion when uniform was discarded and oldest clothes were worn. The Royal Marine Band normally played all day. This picture was taken in 1907 aboard HMS Queen *in Malta.*

Below:
A garden party at Eastney on the occasion of the visit of Imperial Japanese Navy officers in July 1907. The RMA Band Play an appropriate musical programme.

The automobile was in its infancy during the 1900s. These Gunners of the Royal Marine Artillery were amongst the first to sample the delights of driving during a vocational training course at Eastney in 1909. The instructors (centre and right) were civilian.

Below:
Musicians and Band Boys of the Royal Naval School of Music at physical training in the gymnasium at Eastney Barracks on 8 February 1909.

Amateur theatricals have always formed an integral part of 'out-of-hours' activities. This production of Arthur Pinero's 'Dandy Dick' was performed in the Globe Theatre at Eastney in 1910. At this time the cast were mainly officers and their wives. Today amateur dramatics are perpetuated in the annual Royal Naval Theatre Festival. This Globe Theatre was demolished in 1972.

A Church Parade at Deal entering through Jubilee Gate into South Barracks c1910. There are three Barracks at Deal, North, South (or Cavalry) and East Barracks and they date from 1794, but did not come into Admiralty possession until 1869. Music stands can be seen bottom right outside the Officers Mess for use when incidental music was played after the parade.

CHAPTER 7

1914-1918
SLAUGHTER AND TRIUMPH

'During the whole of my service in the Navy I have always felt the utmost confidence in the work of the Royal Marines; whether serving on shore with a Naval Brigade, or afloat manning guns or turrets they have been equally splendid.'

Admiral of the Fleet
Earl Jellicoe of Scapa GCB OM GCVO

When war came in August 1914, there were no such mad schemes as those proposed by Lord Fisher in the first decade of the century. The Great War was one of attrition and logistics. Expertise and skill, as typified in Britain by the B.E.F. (British Expeditionary Force) and the Royal Marines, as with similar such èlite units in all other warring nations, could no longer make an impression. Personal courage, steadfastness, discipline and all the other qualities emphasised in warfare hitherto simply led to greater slaughter of the best young men of the nation. By the second year, all such expertise was reduced to the basic ability to walk across several hundred yards of mud, through barbed wire entanglements, machine-gun fire, artillery barrages and clouds of poison gas to occupy the shell holes of the enemy. In the gigantic mincer of the Western Front the flower of the nation's youth was pulped by uncomprehending generals, seemingly out of their element, and praised by the media who were equally out of touch with reality. If war had been glamorous before, which is doubtful, it was certainly not to be so in the future.

In the midst of such universal carnage all that an élite fighting force, such as the officers and men of the Royal Marines, could do was to show that they could die bravely and well, doing their best amidst overwhelming disaster. It might be supposed that rational thinking would have led the long established Royal Marine Artillery and Royal Marines Light Infantry to act in mutual support of each other as a coherent whole. Nothing of the kind occurred, as units were used piecemeal and in isolation. Thus the Royal Naval Division, the bulk of which vanished forever in the swamps of the Somme and Passchendaele, relied on army artillery support for its forlorn offensives,

A group of Royal Marine Artillerymen filling water-bottles with water provided by local inhabitants in Belgium in 1914.

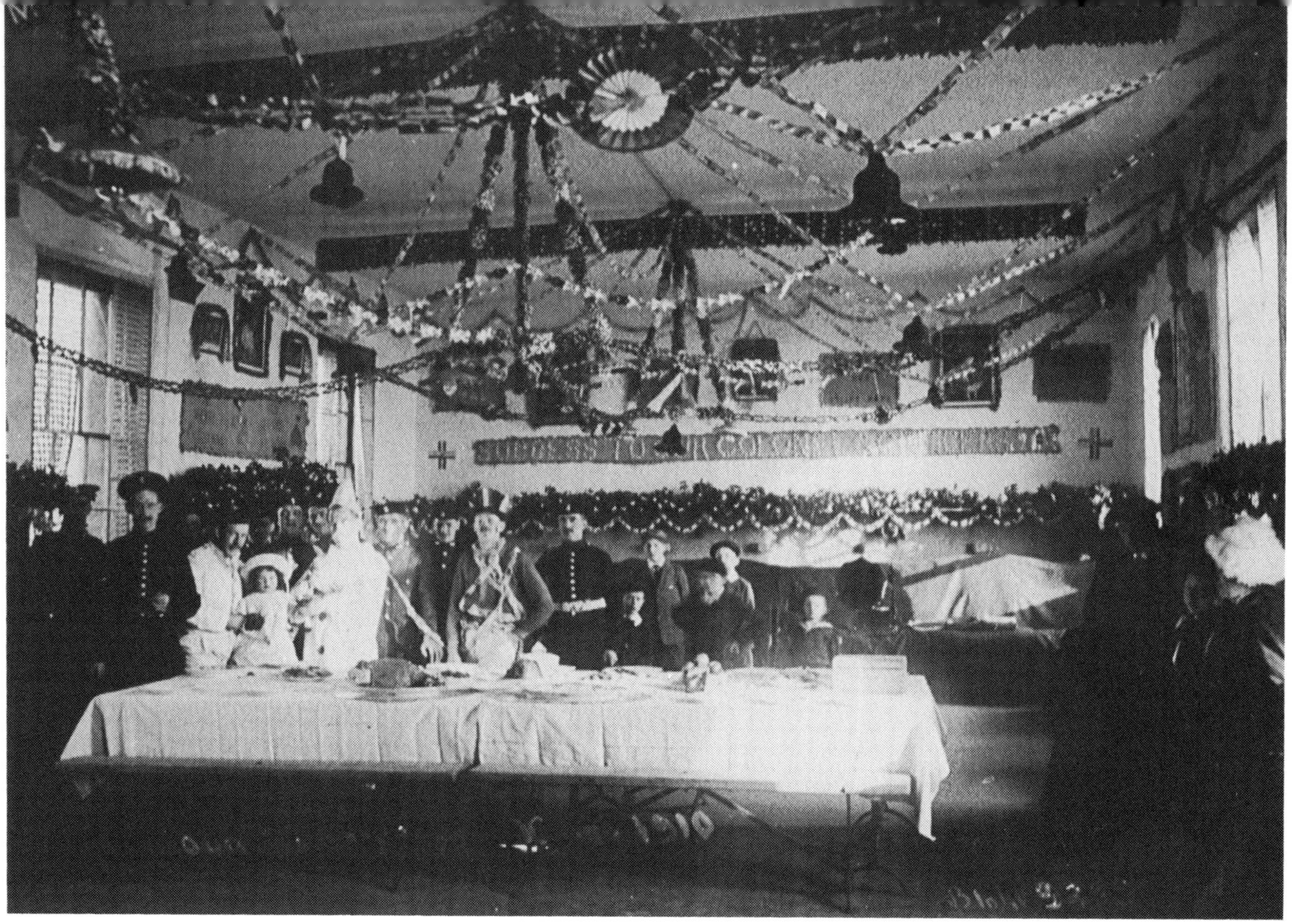

while the great guns manned by the RMA boomed away carving the French and Belgian countryside into water-filled craters to aid army battalions slithering forward into oblivion.

Only at the very beginning of the war, in the period 1914-15, could the Royal Marines ashore make a small recognisable and uniform mark, albeit a small scratch on overall events. Thus on general mobilisation in August 1914, the Royal Marines Brigade was formed with Chatham, Portsmouth, Plymouth and the RMA providing a battalion apiece, in readiness for the war that was confidently predicted to be like all of Britain's long wars '...over by Christmas.'

Of course it was not over by Christmas, but the war of movement on the Western Front was. The great arc of the German army's advance had been held up at the Marne and the 'race to the sea' began. As each army tried to outflank the other they ran out of land, with the Swiss frontier at one end and the English Channel at the other. Just how far along the Channel that front was to be anchored was, of course, vital to Britain. Once the great web of forts around the Belgium fortress of Liege had been battered into submission by the giant German siege guns, then the obvious target for their advance was the port of Antwerp. The German army arrived there on 28 September 1914, and the Scheldt again assumed its vital importance as was long ago recognised by Napoleon.

By 3 October it seemed that the Belgians must soon surrender Antwerp as they were being over-

Christmas is celebrated in an Eastney Barrackroom in 1910 when families were allowed to visit, the only occasion during the year. Such slogans as 'SUCCESS TO OUR COLONIAL COMRADES' were common at this time. Those who served at Eastney until the late 1960s will see there was very little change in barrackroom accommodation in the intervening years.

whelmed by German might, but the youthful First Lord of the Admiralty, Winston Churchill, had one of this ideas (either inspirationally heroic or hideously amateur depending on whether you were to be the sacrifice or not) and proposed to an incredulous Cabinet that he, in person, should lead a force to save the city. They refused his first proposal, thankfully, but accepted the second. The unit that Churchill selected to hold the might of Germany on its own was the Royal Marines Brigade under the command of Brig-Gen A Paris RMA. Hasty and ill-prepared mobilisation had resulted in a large number of fresh young officers and new recruits with just a leavening of veterans. This put paid to the much-vaunted 'flying-column' of pre-war plans. Even so, the Marines were in much better shape than others who were soon to join them.

The Royal Marines Brigade had, in fact, only just arrived at Dunkirk from England after being landed at Ostend on 28 August. It had then been brought back home after one week. About 200 of these

The RMA Anti-Aircraft Brigade was formed 500-strong in 1914 with four batteries each of four 2-pdr pom-pom guns mounted on armoured cars. They first saw action in France in April 1915 in the Nieuport and Dunkirk areas where they formed mobile patrols.

Marines were sent back to Dunkirk to operate armoured cars of the Royal Naval Air Service, the 'motor bandits' as they were christened. On 12 September a freshly-raised battalion from Deal joined the Brigade in place of the RMA and the whole formation sailed for the continent again a week later. On Churchill's orders they duly moved up to Antwerp on 5 October, and here they were joined by the raw boy recruits and reserve veterans of the 1st and 2nd Naval Brigades from England on the 6th.

The Royal Marines took over the shallow and useless Belgian trenches and fought hard under Lt-Colonel C McN Parsons RMLI, acting as brigade commander until the arrival of Col A E Marchant on 6 October. Fortunately they were pulled back just before their lines were pulverised by the German heavy artillery, and under Marchant, fought on in the inner defences. They managed to delay the inevitable by a few days, but then catastrophe overtook them. The withdrawal began on the 7th in the face of enormous pressure. Antwerp fell on 10 October and in the process the bulk of the Naval Division was lost. About 1,500 lost their way and were interned in Holland for the rest of the war, others, about 1,000, were killed or taken prisoner. It was in this action that the Marines suffered badly when the train carrying the Portsmouth battalion was captured by the Germans. Paris managed to extract the rest of his Brigade just in time, but overall it was a disaster and the Marines losses were about 400 men. Churchill tried to justify the sacrifice of these young men so early in the war, but others saw matters very differently. Admiral Richmond, for example, had written of Churchill's scheme *before* the disaster that:

> I don't mind his tuppenny untrained rabble going, but I do strongly object to 2,000 invaluable Marines being sent to be locked up in the fortress and become prisoners-of-war if the place is taken. They are our last reserve. No Board of Admiralty with two-pennyworth of knowledge and backbone would have allowed Marines to be used in such a way...

Admiral Beatty wrote afterwards that Churchill '...must have been mad...' to have thought he could hold Antwerp 'by putting 8,000 half trained troops into it.'

It was obvious that a period of re-organisation was urgently required for the Corps, whose strength at the outbreak of war was about 18,000. A special training establishment was set up at Blandford Forum in Dorset with the intention of concentrating the reformed Naval Division there. The 'Blue' and 'Red' Marines now became 'Khaki' Marines for the duration, and heavy recruitment into the RMLI was initiated to man the specialist units for the Division, in particular signals, transport, engineers, medical teams and a cyclist company. This rejuvenation had hardly got under way when it was abruptly terminated. The Royal Marines Brigade HQ, Chatham and Plymouth battalions were ordered overseas and under the command of Brig-Gen C N Trotman they soon found themselves in the Eastern Mediterranean.

Ever since Turkey entered the war on the side of Germany schemes had been afoot to re-occupy Gallipoli, force the Dardenelles and threaten Constantinople with bombardment to bring about Turkey's early surrender. We had tried this plan before in 1807. The results of the 1915 campaign were exactly the same - nothing finally achieved, only an appalling loss of life.

It was at first thought that bombardments by the fleet to wreck the forts would suffice and, following this up, the two battalions of Royal Marines (about 2,000 men) sent out by Churchill on 6 February, arriving at Lemnos on 23rd. The first of them landed from the bombarding battleships *Irresistible* and *Vengeance* two days later, to complete the work started by the big guns. Trotman's marines were subsequently deployed there on 1 and 4 March, No 4 Company of the Plymouth Battalion landing at Sedd-el-Bahr, while No 3 Company, under Lt-Col G E Matthews managed to take on a whole company of Turks near the Kum Kale battery on the Asiatic shore, losing 44 men, of whom half were killed, before extricating themselves smartly.

Under Brig-Gen Paris, the rest of the Royal Naval Division arrived at Lemnos on 12 March. There followed a lengthy build-up in preparation for a frontal assault. In this the Royal Marines played a prominent part, which was reflected in their casualties. On 25 April the Plymouth Battalion again went ashore as a feint on 'Y' beach to allow troops on 'X' beach to establish themselves. Told they would only be ashore for six hours, Col Matthews' men hung on for thirty, with no support, no orders and no clear sight of the well dug-in Turks who subjected them to merciless fire the whole time. Heavy massacres were taking place on the main Allied landing beaches and all the Marines' reports were totally ignored in the confusion. No 2 company was ordered to re-embark without the CO's knowledge but the survivors of Nos 1, 3 and 4 Companies hung on grimly for another 24 hours and repulsed several heavy attacks. Matthews then took matters into his own hands to prevent complete annhiliation and managed to withdraw the pitiful remnants.

No 4 Company (Capt C A Andrews) of the Plymouth Battalion RMLI land on V Beach at Sedd-el-Bahr on the southern tip of the Gallipoli peninsular near Cape Helles on 4 March 1915. The landing, which was unopposed, acted as a covering force for the demolition parties and to allow aerial reconnaissance for a site for an aerodrome. They later withdrew in good order.

Naval cutters landing men of the Plymouth Battalion RMLI (Lt-Col G E Matthews) on Y Beach on the Gallipoli peninsular on 25 April 1915. The frontal assaults on V and W Beaches met very heavy opposition, but Y Beach was comparatively quiet. The Plymouth Battalion expected to be ashore for only 6 hours but they subsequently had to hold out for 30 hours with little support.

It was soon the turn of Chatham, Deal and Portsmouth battalions to suffer the same fate as Plymouth. Composed of 75% raw recruits, 20% reservists and a mere 5% active service men, they were thrown ashore to relieve the Australians, who had taken a battering at ANZAC Cove on the 28th. Once again they were asked to hold the line for 48 hours, which stretched to 14 days! Taking over the shallow Aussie trenches, they were just in time to meet wave upon wave of Turkish infantry assaults launched by the Turkish commander, Mustafa Kemal, against their tiny enclaves on 1 May. They could not retreat and they were too weak to advance as they had little food or water, and not much ammunition. Still they held out, one platoon of the RMLI for example, was totally cut off but managed to hold their line for four days and three nights.

A stretcher-bearer, L-Cpl W R Parker RMLI, won the Corps a Victoria Cross which might equally have been awarded to scores of Marines in this period. But this does not detract from his heroism when he led a team of volunteer water-carriers through to an isolated trench across a 400 yards stretch of open ground swept by continuous heavy machine-gun fire. It was suicidal and the whole party were obliterated in moments except for Parker who alone reached the trench. Here he helped tend the many wounded and then organised the break-out back to safety of the few survivors. The bulk of the marines had been relieved by the 2nd but were almost immediately ordered back into the line to restore a Turkish breakthrough and to launch a vigorous counter-attack which saved the day. The Portsmouth and Chatham battalions remained in the line until the 12th.

Following this, the Marines moved to Helles on the European side of the Straits under the baleful glare of the enemy on the Achi Baba heights. As the en-trenchments were dug deeper and the barbed wire became thicker, futile assaults followed each other

In the trenches on the Northern Barricade at Cape Helles, Gallipoli in April 1915. Capt D J Gowney (left) uses a periscope to view enemy lines, while Capt C G Farquharson, Adjutant of the Portsmouth Battalion, rests during one of the lulls in the fighting. Capt Gowney had won the DCM as a Corporal at the Peking Legations siege in 1900, was the first Royal Marine Officer ever to be awarded the DSC for service at Antwerp in 1914 and was subsequently Mentioned in Despatches three times during World War I.

A parade of the Royal Marine Artillery Siege Batteries who were sent to East Africa early in 1916. Because of shortages of heavy artillery the Admiralty agreed to send four 4" Mark VII guns and four 12-pdr guns on field mountings. The batteries scored many successes against the Germans in their two year stay.

into this labyrinth. As the long hot summer wore on disease added to their woes. In August they had been so decimated that the pitifully few survivors of Chatham and Deal battalions had to be reorganised as the 1st RMLI; the rest formed the 2nd RMLI. Here they remained wasting away until November and the first icy blasts of winter. Early in January 1916 the commanders had to face the inevitable, and they decided to withdraw. As so often before, the Royal Marines had the doubtful honour of being the rear-guard, the 1st RMLI being evacuated on 8th, while the 2nd RMLI, under General Paris, were the last Allied troops to leave, embarking at 0330hrs on the morning of 9 January. From there they were sent to the Salonika front and remained in that 'open prison' until May, when they sailed for home. Even greater sacrifices were now to be demanded from these same men.

Perhaps the first Royal Marine to engage the Turkish army in The Great War had been Pte Frank Ward RMLI, one of the 'Red' Marines from Forton Barracks Gosport, under Capt J W Snepp RMLI, serving aboard the cruiser *Minerva* when she shelled the fort at the head of the Gulf of Aqaba with 6" shrapnel on 1 November 1914. Later a landing party went ashore and demolished the fort and nearby barracks.

The RMA, as usual, provided guns crews for a wide variety of static coastal defences during the early part of the war, as far apart as Cromarty and Scapa Flow in the north of Scotland, to Capetown and beyond. They also served in the field during the campaigns which led to the conquest of German possessions in East and South-West Africa. In October 1914, 54 officers and men of the RMA with four 4-inch and four 12-pounder guns went out to aid the Union Defence Force returning to the Western Front in 1915. The following year other Royal Marine Artillery men, reinforced by Marines from the East Indies squadron fought in East Africa until 1917.

Since October 1914, the RMA had been reorganised at home in order to provide two artillery brigades for use on the Western Front. A recommendation that three artillery brigades should be formed had been rejected. One unit, 500-strong, fitted out at Eastney in December 1914 in the then novel role of an anti-aircraft brigade. It was fully mobile and equipped with 2-pounder pom-pom close-range guns customised with special sights designed by Col Osmaston. There were four such guns to a battery, each mounted on an armoured truck; four batteries in all were formed. The first two of these reached the Western Front in April 1915, 'B' Battery stationed at Coudekerque shooting down its first enemy aircraft on the 30th of that month. These two were joined by the other pair in the autumn. They mainly operated along the coast from Nieuport to Dunkirk where there was heavy enemy air activity, and also in the Ypres salient. Several of their trucks were destroyed by enemy artillery and the occupants of others were gassed in the salient, but it was the increasing height at which the aircraft flew that brought about the demise of most of them. Eventually their task was taken over by the army, although the Dunkirk battery stayed a Royal Marines unit after re-arming with the long range 3-inch HA gun in 1917. The German air raids on London saw the formation in June 1918 of a second RM battery equipped with the

A group of sailors and marines relax off duty on Ascension Island during World War I. War seemed comparatively remote on this island which was garrisoned by the Royal Marines from 1815 until 1923. Pte W G Blunden is on the right.

3-inch gun for service in England. The two RM batteries were commanded by Major Briscoe and Major Cartwright until after the armistice and were finally disbanded at Eastney on 16 January 1919.

The second unit was equipped with heavy 15-inch howitzers. These were designed by the Coventry Ordnance Works and were meant as counters to the super-heavy weapons used by the Austrians to destroy the Liege and Antwerp defences. Admiral Bacon, a retired Royal Navy gunnery expert, now manager of the Ordnance factory, was given a special commission as a Royal Marine Colonel to supervise the construction of the guns.

Churchill was always impressed by heavy guns, and he ordered twelve such weapons. The emphasis was on mobility and a specially-designed train for each gun had to be built, consisting of eleven tractors and twelve trucks with a team of 88 men to man them. Each gun required one such team and the weapon was designed to be broken down for loading and moving around the front line. Being howitzers, their range was comparatively short, 10,800 yards. They fired a 15-inch shell which had to be hoisted into the breach by means of a sturdy gantry and pulley and chain mechanism. Their short range meant they had

A 15" Howitzer in action on the Western Front of France during the Great War. Twelve of these were in the RMA Howitzer Brigade being mounted on platforms. Rear Admiral Bacon, manager of the Coventry Ordnance Works, who took them into action, was granted a commission in the Royal Marines as a Colonel 2nd Commandant. The projectile weight was 1,400lbs and had a range of 10,800 yards.

to be used close to the front line, which made them extremely vulnerable to counter-battery work, but the risk had to be accepted.

The full Howitzer Brigade of the RMA totalled some 1,000 officers and men, but it seldom fought as a single unit. More commonly it operated in single gun sites scattered along the front. The first pair of the eleven gun that actually saw service in France, arrived in February 1915 and were first in action at Locre on 6 March. Admiral Bacon was succeeded for operations in the field by Major F W Lumsden RMA, who commanded the brigade between April and July 1915. The latter subsequently went on to hold a number of army staff appointments including command of the 17th Highland Light Infantry, where he won the Victoria Cross organising the withdrawal of captured German guns under heavy enemy fire. Later he commanded the army's 14th Infantry Brigade and when he fell in action on 3 June 1918, he had a gallantry CB and three DSOs in addition to his VC.

Bacon had other Royal Marines connections for he was appointed to be second-in-command of the famous Dover Patrol, which included in its defences several long-range guns manned by Marines and seamen as the RM Siege Regiment. A single 12-inch gun (*Dominion*) under the command of Capt H Peck RMA and four 9.2-inch guns (including *Eastney*) commanded by Capt J H Hollingsworth RMA were shipped over to Adinkerke and Groenendijk respectively in December 1915 for counter-battery work. Their main role was long-range duelling with their German equivalents, the 11-inch naval guns of the *Tirpitz* battery across the channel at Ostend.

These units were much expanded in 1917 in readiness for the Passchendaele attacks when the Royal Marines took over the whole force under the title of RM Siege Train. They added two more 12-inch, four more 9.2-inch and six 7.5- inch naval guns which were carried in fortress mountings, served by 500 officers and men. From June 1917 they fought long-range battles at ranges up to 32,000 yards (18 miles) with the German 15-inch *Leugenboom* with neither side being incapacitated. All these great weapons were dismantled at the end of the war and the unit disbanded at Eastney in March 1919.

In the meantime, the RMA Howitzer Brigade was employed in almost every one of the abortive battles leading to the final Armistice. Another pair of 12-inch guns had been shipped over by April and ten of them were in action during the great Somme offensives of 1916. They also served in a similar capacity at the Passchendaele in 1917. Counter-battery fire was particularly potent in the campaign and several of the guns crews took heavy casualties, in all losing 35 killed, 97 wounded and 69 gassed. They were eventually merged into the 1st and 2nd Royal Marine Siege Regiments and re-equipped with the standard army 12-inch howitzers. In this role they saw out the final battles of the Western Front, even utilizing captured German 4.2-inch guns during the final advances of 1918.

We must now return to the two battalions of Royal Marines serving with the Naval Division. This division had been transhipped via Marseilles to the Western Front in May 1916 and it had been renamed the 63rd (Royal Naval) Division, and its two Brigades the 188th and 189th Infantry Brigades. In came army brigadiers to command them while a third brigade, the 190th, consisted of army units, as did the supporting artillery field brigades for the Division. Machine-gun companies were added, with recruitment for the

A RMA petrol tractor tows the barrel of a 7.5" Mk 3 gun. These Foster-Daimler tractors were first designed in 1912, were of 105 HP and had rear wheels with a diameter of 8'. It needed eight tractors to tow one complete gun when ammunition was moved as well.

RMLI remaining at Deal. Thus the Royal Marines content of the force was considerably watered down consisting only of the 1st and 2nd RMLI, who formed part of the 188th, along with Howe and Anson battalions. The 189th consisted of Hood, Drake, Hawke and Nelson battalions and RMLI officers commanded the latter three for a time. In addition Brig-Gen Trotman commanded the 190th.

A heavy blow to morale came with the wounding of General Paris while out on a reconnaissance on 13 October. He was a much respected figure who, as a field soldier, understood the special nature of his command. His replacement, an army general, cared nothing for naval traditions and turned them into just another trench-warfare outfit to be ground into the mud and wire as the early promise of the July offensives foundered in the mire of November. By this time, there was no hope of any great break-through and divisions were fed to the meat-grinder merely as a means of attrition. The marines battalions were placed into the line for long periods and their numbers were worn down by continual shelling and patrolling. Even during their brief reserve periods they found life was no sinecure, as they were constantly being called forward to fill in gaps in the line.

The day of reckoning came at the Battle of the Ancre when they finally 'went over the top' in four waves, within seven minutes of each other, in the early hours of 13 November 1916. At the time each battalion was barely 400 strong, instead of the complement of over 700. They scrambled across the 200 yards of rain-drenched mire that had been turned into a filthy morass by continual shelling, where they found facing them four powerful lines of German trenches around Beaumont Hamel. The remnants managed to reach the third line after much bitter hand-to-hand fighting. The RMLI made flank attacks to secure ground while their comrades eventually took Beaucourt. For two days the assault was kept up and reserves fed in, but when the marines were relieved on the 15th the 1st Battalion was reduced to 140 officers and men and the 2nd Bn to 152. Three of the four RMLI colonels lay dead and the other was wounded.

Reinforced and re-trained, the RMLI battalion took part in three further offensives in 1917. They were in the attacks on Miraumont on 17/18 February, on Gavrelle on 23/24 April, where they took particularly heavy casualties, and were at Arleux on 28/29 April. All these nibbles were but a prelude to the great offensive launched by General Haig at Passchendaele that autumn. One example of the difficulties of fighting in the quagmire of the Western Front is epitomised in the RMLI attack on Peddebeek

A 12" Mounted Railway gun of the RMA Siege Regiment at Woestea, Belgium on 23 August 1917.

One of the earliest three-ton vehicles takes Boy Buglers on an outing during World War I.

Brook. Months of heavy shelling, coupled with continuous rain the night before, made a mockery of the detailed maps which had been issued to the commanders. The battlefield was just a sea of churned-up, glutinous mud, often thigh-deep into which they were moved forward at 0504hrs on the morning of 26 October. The rain continued all day as Lt-Col Harold Ozanne RMLI led the 1st Battalion out in three waves. It was yet another brave and heroic slaughter. Lt P Ligertwood RMLI, for example, led 'A' Company to their objective despite being wounded three times. Such courage is hard to imagine, but he still managed to urge the survivors on before being wounded for the fourth time, this time fatally. Counter-attacks in the sludge forced them back from even the most modest gains, and in the end the two battalions had gained no more than 350 yards, losing 700 officers and men in the process. 'No troops could have had to face worse conditions of mud' wrote General Gough, one of the few senior officers to see what the front was actually like, 'No troops could have done more ...' Another general, visiting the scene after the battle, is reported to have broken down and wept. 'Did we send men to fight in this?' he asked. They had indeed.....and they were to continue to do so.

A German winter offensive was mounted across snow-covered ground against Welch Ridge at Cambrai on 30/31 December in order to regain what they had lost. Only the decimated ranks of the RMLI were there to meet them. The Germans penetrated the finely-held lines, but the Marines stuck it out for a long time only giving way under intense and overwhelming pressure. Another winter's stalemate was endured and the RMLI, now reduced to two small battalions, took the brunt of the German spring offensive which burst like a bombshell on 21 March 1918. Mustard gas took a heavy toll, but they conducted a fighting retreat for five days. In the end they turned and counter-attacked Aveluy Wood on 6 April with great fortitude and ultimate success.

Finally triumphant they stopped the enemy advance, but it was a pitifully reduced force that emerged victorious from this bruising encounter. There were not sufficient reserves to make up the numbers and the two battalions were merged into one for the rest of the war under Lt-Col E K Fletcher RMLI. As the 1st RMLI they took part in some of the great attacks that marked the final phase of the war on the Western Front, when the German armies were steadily being rolled back. They took part in the battle for Albert on 21/23 August 1918, the colonel being wounded and gassed in this encounter. He never really recovered and died four years later in Deal. By September the RMLI were engaged at the breaking of the Hindenberg Line at the Drocourt/Queant Switch and they fought over the Canal du Nord and the Canal de l'Escaut, before taking part in the final

battle of Cambrai before the Armistice ended the four years of slaughter. Typically, on 11 November, the Royal Marines found themselves close to Mons, where the first British battles had been fought in 1914. All that death had gone full circle.

Very few of the original battalions came through those four ghastly years; the losses suffered, as Churchill commented, were '...of so awful a character as to sweep away three or four times over the original personnel.' Some 6,000 Royal Marines officers and men fell in the Great War, the greatest losses ever to be suffered by the Corps in a single conflict. Among the few who did survive to ponder some terrible memories of many friends who would never come back, was Capt H T Burton RMLI, the Quartermaster of the original Deal Battalion, who was present at the final parade attended by the Prince of Wales, before the 1st Bn RMLI was disbanded in June 1919.

But it was not only on land that the Royal Marines served with courage, distinction and heavy loss. They took their fair share of the fighting at sea and in the air as well. Flying was in its infancy when war broke out, but the Corps was already well represented among the early pioneers of aviation. Two Marine lieutenants earned their 'wings' as early as 1911. Lt E L Gerrard RMLI had been selected for duty with

During the first Battle of the Somme on the Western Front in 1918, 1st & 2nd Battalions RMLI formed part of the Royal Naval Division. The German offensive of 21 March broke through on the right and the Royal Marines found themselves defending the Green Line near the Metz switch. Here some of the marines rest on the banks of the canal after the battle of St Quentin.

No 1 Rigid Naval Airship (popularly known as the *Mayfly*) in 1910, and in the following year was one of the first four officers to train as naval pilots. Another Lt G V Wildman-Lushington RMA, had been selected for the course but went sick, and Lt R Gordon RMLI became the second RM pilot in December 1911. Subsequently Gerrard became one of the first instructors when the Central Flying School was formed on 19 June 1912. Lushington became Winston Churchill's personal flying instructor a year later, but unfortunately was subsequently killed in a flying accident. Lt C H Collet RMA led the first air attack on Zeppelin sheds in Germany on 22 September 1914, the first ever bombing raid, for which he was awarded the DSO. He, too, was later killed in a flying accident at Imbros during the Gallipoli campaign. These veterans were joined by 11 other marine officers as

Major E L Gerrard RMLI, the first Royal Marine to qualify as a naval pilot, was awarded his 'wings' on 9 May 1911. In 1912 he became one of the first instructors at the Central Flying School and is seen in the cockpit of a Henry Farman at Upavon. It was the practise in those days to land in the grounds of large country houses to attend garden parties. On 31 July 1911 Major Gerrard is recorded as having 'given a flip' to Miss Kerr, Lady-in-Waiting to Princess Louise of Battenburg. He was later awarded the DSO at Gallipoli and transferred to the RAF on its formation on 1 April 1919.

the British naval and army air arms expanded. During the war they won between them three CMGs, one CBE, one CB, nine DSOs, one OBE and two AFCs. Royal Marine aviators were thus well to the fore in this new venture. When the Royal Naval Air Service and the Royal Flying Corps merged to become the Royal Air Force on 1 April 1918 ten Royal Marines officers transferred to the new Service.

At sea, the Royal Marines continued in their traditional role as gunners. About 10,000 from a total Corps strength of 18,360 officers and men, were serving afloat at the end of August 1914. The enormous mushrooming of personnel during the war forced the Corps to enlist short service officers and men who signed on for 'the duration'. Initially this caused many frowns with talk of dilution, but the majority of men so recruited were soon fired with the zeal and eagerness to serve and thus made excellent Marines. At the time of the armistice, the Corps had grown to a strength of 55,603, which foreshadowed the enormous increase during World War II to a peak of over 78,000.

The concentration of the front line strength, of what was to become Admiral Jellicoe's Grand Fleet, and the competent commissioning of the reserve fleet was most timely. As the tension grew, the order for the latter to disperse was cancelled. While the great battleships and battle-cruisers of the Grand Fleet sailed unmolested up Channel to their wartime base at Scapa Flow in the Orkney Islands, the bulk of the older ships remained concentrated at Spithead. Mobilisation telegrams had already been despatched on 2 August. Lt-Col G R Hawkins gives an excellent account of how the Marines transferred from peace to war in those early weeks. He was at the time serving aboard HMS *Minerva* and wrote later:

> The Royal Marines were all Red Marines from Forton Division. Almost half were reservists who, in

civilian life were postmen, policemen, firemen, etc. There were three Bisley shots in the detachment, and two footballers who played in the famous Lillywhite team and that won the Army Cup and the Amateur Cup in the same year, 'Pinky' Yates and 'Jock' White, who died of black water fever in German East Africa in 1917. Most of the detachment had two or three good conduct badges and several had the long service and good conduct medal.

On 4 August, whilst we were at "Quarters Clean Guns", the Bosun's Mate interrupted proceedings when he piped "D'ye hear there"...then after a short pause for all hands to stop work and listen..."we are now at war with Germany. Hands carry on with your work."' *Minerva* then got up steam and proceeded to her war station. An order was received that to prevent fire spreading, all wood, not essential for war, was to be disposed of. It was a very sad sight to see the Wardroom piano and the Ship's Company piano side by side slowly drifting astern as we steamed away from Spithead.

There were many shocks at sea. Amongst the first was the vulnerability of the older warships, which had huge crews but limited uses, and the new underwater menaces of submarine and mine. Early losses included the battleships *Audacious* and *Formidable* and the cruisers *Aboukir, Cressy*, *Hawke, Hogue* and *Pathfinder*. All except the first, which was mined, were torpedoed by German U-boats in the first months of the war and sank with heavy loss of life, including many irreplaceable Royal Marines. Equally unexpected were the total loss by mysterious explosions of the battleship *Bulwark* on 26 November 1914 at Sheerness and the armoured cruiser *Natal*, which caused heavy casualties, without any chance of striking back at the enemy. Nor did the early sea battles produce the long-awaited victories over the German foe. Instead, at Coronel two more big cruisers, *Good Hope* and *Monmouth*, were out-ranged and sent to the bottom in hopeless fights. Yet more marines died at the Dardenelles when the old battleships *Irresistible, Goliath, Majestic, Ocean* and *Triumph* went down. Seldom had the Royal Navy suffered so many disasters in such a short space of time. The failure to bring the enemy to book, either by intercepting the *Goeben* in the Mediterranean or the German battle-cruisers, save for the old and small *Blücher* at Dogger Bank, did little to balance the scales, although the Falklands battle on 8 December 1914 was a solitary success on which to hang hope for the future.

That hope took a long time to come to fruition. Years of searching the grim North Sea and promising vengeance '...if only they would come out and fight...' finally produced one unique opportunity off Jutland on 31 May 1916, when, in the murky afternoon, the Grand Fleet cut off the ships of the German admirals Scheer and Hipper. 132 Royal Marine officers and 5,832 men were serving in the ships of the Grand Fleet that day. Alas, although Jellicoe brilliantly

The funeral cortege of Capt G V Wildman-Lushington RMA who was killed in a flying accident at Eastchurch on 2 December 1913. He was buried with full military honours in Christchurch Cemetery, Portsdown Hill, Portsmouth. Capt Wildman-Lushington was one of the first Royal Marine officers to become a pilot in 1912, and in 1913 gave dual instruction to Winston Churchill, then First Lord of the Admiralty. His brother (G E), also an aviator, became a Major General and was Chief of Combined Operations from 1947 to 1950.

Capt Robert Gordon RMLI (later to become an Air Commodore RAF) takes off from the Drill Field at the Depot at Deal in a Short S27 Pusher Biplane in 1912. He was one of the earliest Royal Marine pilots, receiving his wings on 7 December 1911. He was awarded the DSO for operations against the Konigsberg *in the Rufigi River in East Africa in 1915.*

crossed the enemy's 'T', the longed-for annihilation did not come to pass. Instead the Germans escaped and worse, in the process, sank five more great ships - the battle-cruisers *Queen Mary, Indefatigable* and *Invincible* and the armoured cruisers *Defence* and *Warrior*. Most of these enormous vessels blew up instantly, taking with them hundreds of seamen and Royal Marines. Others died in ships that were badly hit, the most memorable example being that of Admiral Beatty's flagship, the battle-cruiser *Lion*, which would have been lost also but for the supreme bravery of one Royal Marine.

The *Lion* was hit on Q (the midships or Royal Marine) turret by an 11-inch shell. Such hits had already accounted for two of her sister ships and *Lion* seemed almost certain to join them. The roof of the turret was blown off killing or fatally wounding everyone inside. Major F J W Harvey RMLI had both legs blown off and was dying, but he managed to order the main 13.5-inch magazine to be flooded. It was only just in time for, as marines from the secondary 4-inch battery fought to get the fire under control, a cordite block fell back from one of the gun breeches and ignited further charges in the crippled turret. In an instant the flash passed down the ammunition trunking wiping out the magazine and shell-room crews, only to be blocked by the closed doors of the flooded magazine. Such dying devotion to duty which on this occasion was rewarded with a posthumous VC, was the hallmark of Royal Marines afloat. At Jutland 538 died and 150 were wounded.

The British fleets were never to get another chance of bringing the German fleet to battle in two and a half years of weary patrolling in fair weather and foul, coupled with long boring spells at Scapa. Such was the lot of Royal Marines serving with the main fleet. The one big consolation for all their losses and unrewarded effort came with the surrender of the entire German High Seas Fleet, the second most powerful in the world, at the end of 1918. It might not have been as spectacular an end of enemy seapower as Trafalgar, but the anchoring of those shamed 'dreadnoughts' under the guns of the Royal Navy in Scapa Flow meant that the Kaiser's challenge to British sea supremacy had ended just as finally and ignominiously as had Napoleon's and Philip of Spain's before him.

One stirring final action gave the Royal Marines the chance to shine in an operation that should have been typical of their work as sea-soldiers in World War I. The attempted blockade of Zeebrugge on St George's Day 1918 gave the nation a great lift to its dwindling war-time morale. Although not completely successful, it featured great gallantry on the part of all those present. The operation also showed what could have been done elsewhere and earlier, had imagination and planning in high places overcome the Admiralty's inherent inertia.

The aim of the operation was to land a storming force from the old cruiser *Vindictive*, especially con-

The upper deck of HMS Vindictive *showing the specially constructed brows for landing on the Zeebrugge Mole. All except two of these gangways were destroyed or damaged on the approach but the marines of the 4th Battalion and the seamen landing parties scrambled across these. This photograph was taken by Lt C N B Rigby RMA who was killed in the foretop and subsequently developed by Lt H Gardner RMA.*

On return to Dover after the epic raid on Zeebrugge by 4th Battalion RM on 23 April 1918, two naval ratings on the upper deck of HMS Vindictive *hold up a Royal Marine greatcoat riddled with holes.*

verted to carry close-range weapons and equipped with landing 'brows' or gangways. These were to be lowered to gain access on to the mile-long Mole from the seaward side at the harbour mouth. Royal Marines were to land across these gangways, storm the German coastal defences and allow the blockships unhindered entry into the narrow channel that the U-boats used for their passage from their base at Bruges, where they would be scuttled. A similar operation was to be launched at Ostend simultaneously. The driving force and thinking behind this scheme was the dynamic, erratic genius Admiral Sir Roger Keyes. Naturally he chose the Royal Marines for the assault force.

The unit selected was the 4th Battalion, which had originally been formed in 1917 to help police Ireland after the rebellion and had subsequently been sent to the Depot at Deal to disband. However, in January 1918, they re-formed with three RMLI companies, one each from Chatham, Portsmouth and Plymouth with RMA support, and a very much more worthwhile mission awaited them. They were reinforced by 134 officers and men from the Grand

When HM King George V visited Deal on 7 March 1918, he posed with the officers of the Depot (Col Comdt H S N White) and the 4th Battalion (Lt-Col B N Elliot RMLI) specially raised for the Zeebrugge raid. To commemorate his visit he decreed that the senior recruit squad should be styled the 'King's Squad' and the best recruit awarded the 'King's Badge'.

Fleet, all volunteers eager to escape from the monotony of Scapa Flow and the North Sea gales.

It was at Deal on 7 March that the battalion was inspected by King George V, their Colonel-in-Chief, who also witnessed recruit training at the Depot. After his visit he wrote to the Colonel Commandant, H S N White,

> ...For some 250 years the Royal Marines have rendered splendid service to the country, and during this great war they have the unique record of having taken part in all Naval actions, and of being employed in every theatre of war. Everywhere your Corps has added fresh glory to its record, and never has your name stood higher than today....I am prouder than ever to be your Colonel-in-Chief.

To commemorate his visit it was thenceforth decreed that the senior recruit squad should bear the title 'King's Squad' and the best recruit awarded the 'King's Badge'. Down to the present day, to be a 'King's Badgeman' and wear the personal cypher of King George V is one of the proudest achievements for any young marine. Even if he is commissioned, he still continues to wear the badge and indeed, one subsequent major-general, E G D Pounds, wore his badge with pride.

But back to Deal, where the 700-strong 4th Battalion under the command of Lt-Col B N Elliot RMLI sailed in eager anticipation on the night of 22 April, a previous sailing on the 11th having been aborted by bad weather. It was not until they were at sea that they were told that their training at Kingsdown for an attack on a fortified post behind enemy lines in France was but a cover story. The *Vindictive* actually got to within 300 yards of the mole before the smoke that had covered the approach was blown away exposing her to the many German guns ranged there. Two German destroyers berthed alongside inboard fired their main guns across the mole into the British ship at point-blank range. Shot and shell tore into the old hull and upperworks, shredding them and punching great holes in her superstructure. Many of the landing party, who were assembled on the upper deck, were killed in those frightful moments, including Elliot.

Up in the *Vindictive's* fighting foretop, Sgt Norman Finch RMA, the sole survivor of his gun's crew, and himself badly wounded, continued to keep his machine-gun in action, which subsequently earned him a balloted Victoria Cross. All except two of the landing brows were destroyed or disabled as Capt A F B Carpenter RN coolly manoeuvred his ship alongside, some 300 yards further down the mole than he had planned. Despite this the survivors dashed across the remaining gangways, the tide being lower than anticipated, in the face of murderous enemy fire. Led by Major B G Weller of Chatham Company, who had won the ballot to lead the attack, Portsmouth and Plymouth Companies followed them ashore clambering atop the sea wall and then dropping 16ft down to the mole proper before beginning a steady advance.

A second VC was won by the Corps that night, again awarded by a subsequent ballot amongst survivors of the 4th Battalion: indeed they earned a dozen altogether. Capt Edward Bamford RMLI of the Portsmouth Company, who had earlier been awarded a DSO at the great battle of Jutland, earned his during an advance down the mole when he showed '...a complete disregard of danger...' in taking several objectives before the recall was sounded. The Plymouth Company, led by Lt C R W Lamplough, were well established in one of the dock sheds before joining the advance in the face of terrifyingly fierce fire. 12 Platoon, under Lt G Underhill, also acted with the utmost bravery when they were sent to the aid of a naval storming party assaulting the German gun positions. Of the 700 Marines who had embarked, no less than 104 died and 206 were wounded in that one memorable hour they were alongside. Yet somehow the wounded were brought back and they sailed away when the *Vindictive* pulled out at 0100hrs. The three blockships, the *Iphigenia, Intrepid* and *Thetis* were sunk in the channel, but the operation did not fully succeed as the Germany were able to dredge a channel around them. However the uplift the news of the attack gave the nation more than justified this final gallant attempt on St George's Day to 'Give the Dragon's tail a damned good twist!' To honour the memory of those who died at Zeebrugge, it was directed that no future Royal Marine battalion would ever be numbered the 4th.

At the end of World War I, the German Fleet were escorted to Scapa Flow, the naval base in northern Scotland where they were anchored. On 21 June 1919 a secret German order was relayed to scuttle the German ships. A British boarding party prepares to board a sinking German destroyer.

Royal Marines have featured on a number of cigarette cards, including Wills, Players, Lambert & Butlers, Ogdens, Salmon & Glucksteins, Churchmans, etc. Whilst these cards are often good sources of uniform information, the two illustrated right have a number of errors.

A hero's welcome for Sgt N A Finch VC when he arrived back at Eastney on 31 July 1918 after winning the Victoria Cross by ballot at Zeebrugge on St George's Day. Details of the award and citation are in Appendix C.

A Peerless Armoured Car. These were used for Internal Security duties in India and Ireland from 1919. It had an Austin chassis and Peerless body (which did not always fit!) was powered by a 40 HP engine, had a range of 96 miles, carried a crew of four and mounted two Hotchkiss guns.

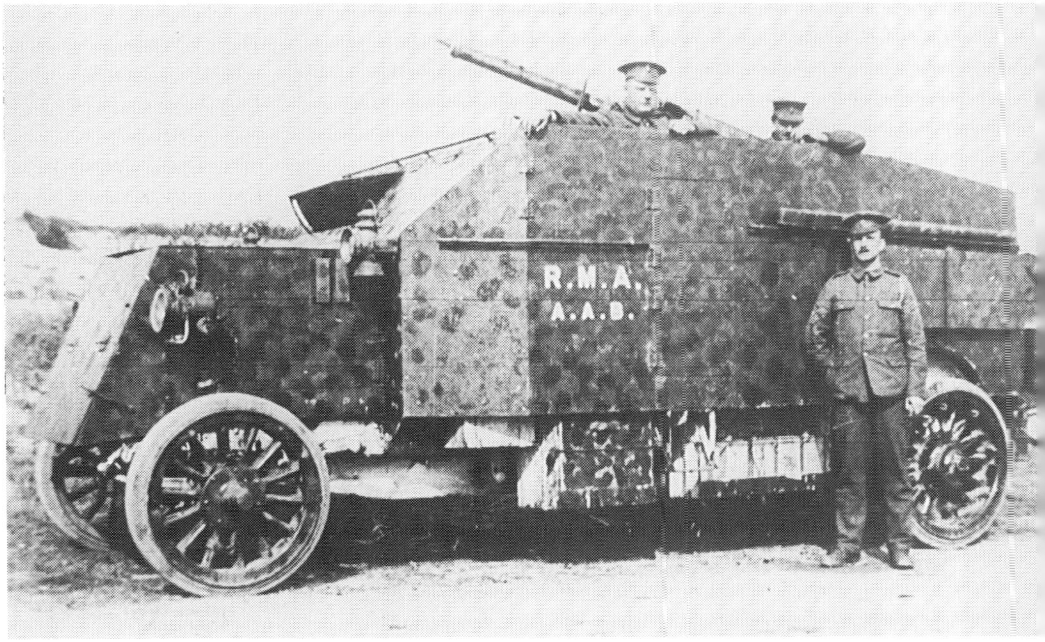

NORTH
RUSSIA

CHAPTER 8

1919-1938

RETRENCHMENT AND RECALL

'The Marines stand for Drill, Discipline and Duty; three words that are not over popular at the present time ...but they commit the cardinal crime of not advertising themselves.'

Admiral Sir Richard Phillimore.
Lecture at RUSI 16 November 1927

It soon became abundantly clear that the 'War to end all Wars' had not done so. In fact it took longer to tidy up the many loose ends left by the Armistice than it did to fight the war itself. Those loose ends involved the occupation of Germany; the campaigns to aid the 'White' Russian forces in their efforts to thwart the Red Army taking over completely; the continuing problems in Ireland leading up to partition; the dealings with the 'Young Turks' and the complications caused by Greek designs on that country, amongst others. All these events saw the much-reduced Royal Marines in action. There were also many tensions in these immediate post-war years due to the retention of some short-service men beyond the duration, and a rare staining of the escutcheon. Finally there came the traumatic disaster of the Washington Naval Conference in which Britain's age-old naval superiority was sacrificed by the politicians in an effort to gain short term goodwill with the Americans. The huge reduction and run-down of the fleet, inevitable after a war, in turn led to a crisis in the Corps, whose future was once more closely questioned and hung in the balance.

In 1917 the battleship *Glory* and the cruiser *Iphegenia* had been sent to the northern Russian port of Murmansk, under the command of Rear-Admiral T W Kemp, to protect the vast amount of Allied war material stockpiled there and at Archangel including 200,000 tons of war supplies and 300,000 tons of coal, in case of the defection of the new Bolshevik Government and an attack by the German army

A group of Royal Marines in North Russia in 1919 wearing dark blue greatcoats and Broderick caps. Note the difference between the cap badges of the RMA (bursting grenades) and the RMLI (the Globe & Laurel and bugle horn). These marines are from HMS Southampton *at Murmansk.*

At an Eastney Open Day in 1919, a Royal Marine artilleryman proudly displays one of the Lewis Guns salvaged from HMS Vindictive *after the raid on Zeebrugge the previous year. It was with one of these guns that Sgt Norman Finch VC fought so gallantly.*

through Finland. The old *Glory* had some of her casement guns suppressed and the extra deck space was used to carry a larger-than-normal complement of Royal Marines for garrison duties. In March 1918, as the situation worsened, more reinforcements including the armoured cruiser *Cochrane* joined the squadron. By April this passive policy had become one of active intervention. The Marines were landed from the ships and soon saw action against the Finnish probes near the border.

At the beginning of May definite reports were received of a pro-German Finnish army advancing on Pechenga to the north-west of Murmansk, while another attack up the rail link through Kem and Kandalksha on the White Sea was also threatened. On 3 May, therefore, a Royal Marines Field Force of 360 men was sent out in the form of a Company of RMLI supported by a Field Battery of the RMA with a Lewis gun section attached. They trained themselves as ski troops to meet the local conditions that winter. Further to-ing and fro-ing in Allied Govern-

When HM King George V visited Dover in 1920, Royal Marines from the Depot lined part of the route.

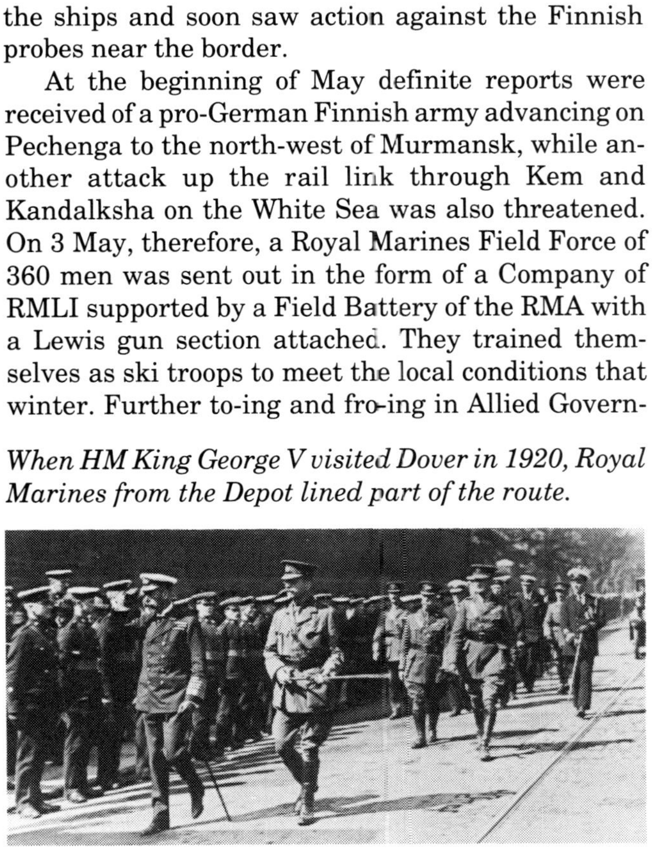

'What have you got to say?' A contemporary sketch by retired Superintendent Clerk Fowler in the 1920s. It depicts an Orderly Room at Chatham and clearly recognisable are the Colonel (left), the Adjutant (with charge sheet), the RSM (with sword) and the escort wearing the Broderick cap.

The first King's Squad ever to take part in the Royal Tournament in 1919, under their instructor Sgt-Major G W Chant, firing a feu de joie. *In the Royal box are HM King George V, HM Queen Mary and Marshall Joffrè. Officers at the gate include Major W K Garnier, Major A R Chater and Lt E J Woodington.*

Summer outings were always a feature of Sergeants Mess life between the wars. Complete with uniformed chauffeur, SNCOs and their wives leave Eastney for a day trip in 1920.

Devon has been used as a major Corps training area for most of the twentieth century. After 20 miles a Field Training class encounter an automobile on one of the narrow Dartmoor bridges during a route march in the 1920s.

mental circles followed as the collapse of Germany saw the reduction of one threat, but an increase in the 'Red' menace. It also saw the Marines anti-German role change, imperceptibly and slowly, into an anti-Bolshevik role, which was not what the Allied troops had originally been sent for.

The first clashes took place in the Spring of 1919 when the Red Army launched an offensive against Archangel, which was hastily reinforced from Murmansk. Under the command of army Col Lecky, this Royal Marine force saw some fighting at Urozeroy on 3 and 4 May, and they also attacked Maselskaya. Soon afterwards they were replaced by a fresh Royal Marine unit from England. The British Government continued to vaccilate, first considering a withdrawal in June, then changing its mind and suggesting an offensive from the Lake Onega region to link up with the White Russians, then again changing its plans. Finally it was decided to hold the Onega area while the White Russians could be trained to defend the area themselves once the Allies had pulled out. This meant aggressive action to keep the area free. There had already been dissatisfaction and the refusal of duties by the Americans, Italians and some Finnish units, many of whom were sympathetic to the Bolshevik cause. However the Royal Marines battalion was ordered to north Russia to support the 'new' policy.

Here we find seeds of grave trouble for the unit concerned, the 6th Battalion, commanded by Lt-Col A de W Kitcat RMLI, which had been shipped out to north Russia instead of being sent, as expected, to police the plebiscite in Schleswig-Holstein in Germany. The battalion had been formed in the summer of 1919 especially for this duty, but had been diverted at short notice. The men were already disgruntled by the unequal form of demobilisation being practised at home. There were many very young recruits among them and they expected to return home after the plebiscite. In the main they were not volunteers for Russia as many of the other units fighting there were, and when they were sent by train from Murmansk down to the Lake Onega region among other mutinous units, it proved a terrible mistake.

In their first engagement with the Reds on 24 June 1919, at Ussuna, under Major Ridings, HQ Company was unfortunately fired on by 'D' Company. Then 'B' & 'C' Companies were caught in an ambush and badly mauled. These misfortunes made an already tense situation far worse and many of the men became demoralised. As a result the best part of two Companies of the battalion returned to camp and refused duties. Courts-Martial were held, the proceedings and results of which have still not been released 70 years later. As a result, 90 Marines were found guilty of whom 13 were sentenced to death, while others were given terms of up to 5 years imprisonment. However wiser counsels prevailed and the death sentences were all commuted by the General Officer Commanding-in-Chief, substituting five years penal servitude when all the sentences were reviewed. This all caused quite a furore in the British press, which the government tried to hush up in vain. The disgraced 6th Battalion continued to serve in the region until the withdrawal of all British troops in September, but the shame remained. There has never been a 6th Battalion since.

The commandeered gunboat named 'Kent' after the cruiser from which the Royal Marines who manned her came, on the Kama River in 1919. 12-pdr guns were mounted with a great deal of improvisation. The other commandeered vessel, a barge nicknamed 'Suffolk', mounted a naval 6" gun. Both saw action in support of the White Russians against the Bolsheviks.

It was not until 20 December 1919 that the First Lord of the Admiralty, Lord Long, was able to tell the House of Commons about the review of the Court-Martial sentences, most of which had been reduced. The real blame lay in the British government pursuing warlike aims against a country and government with whom we had no quarrel, other than the political one of repugnance of the British ruling

classes towards the establishment of a classless state. Their attitude was similar to that shown during the French revolutionary period 120 years earlier. It is not surprising that many of the rank-and-file saw no virtue in this attitude.

This was not the only intervention by Royal Marines in the Russian Civil War. A small mobile group, known as 'Dunster Force', operated for a time across the Persian border to secure the Baku oilfields in the Russian Caucasus. The force included 160 RMA and RMLI who, with two 6-inch guns, garrisoned Petovsk for a time.

Further to the east across the Russian land mass 50 Royal Marines were landed at Vladivostock from the cruiser *Suffolk* in April 1919 to restore order there, but were withdrawn a few days later. However, that did not mark the end of their intervention, for a force under the command of Capt T H Jameson RMLI, with marines from the cruiser *Kent* also, formed an armoured train of 24 trucks and carriages, with living accommodation, provision cars and ammunition wagons with a 6-inch gun and four 12-pounders to fight the 'Reds' on the Kama river. They moved across

The Royal Marines detachment of HMS Kent *and HMS* Suffolk *with the 6" gun which they loaded on a train and transported across Siberia.*

the trans-Siberian railway in support of the Czech army's advance westward, and supported its subsequent withdrawal.

On the Volga a small party of Marines from a battalion of 500 officers and men which had landed on their old battleground of Sebastapol in the Crimea in November 1918 in order to evict a German army of occupation, found itself fighting with the White Russian army on the Volga front in the vicinity of a steel town. The Soviet commander defeated the Whites and drove them back in disorder to the Crimea once more. The town was renamed after him: it was to become renowned in a later war as Stalingrad. Seldom have the Marines found themselves fighting so far from any ocean as those caught up in the heart of Russia in 1918-19.

Yet other Marines were caught up in weird and unplanned situations. A force of 45 found themselves on the bridges of Budapest as the only buffers between two warring armies, those of the newly formed states of Hungary and Rumania. Others had to suppress a mutiny in the far-flung island of Singapore, and yet others occupied the Turkish forts at Gallipoli, at long last, during the peace talks which commenced in 1918.

Some of the 50 Royal Marines from HMS Kent *and* Suffolk, *under Capt T H Jameson RMLI, who had been landed at Vladivostock in April 1919 are pictured at Brammboctoks in Siberia. They formed an armoured train with a 6" naval gun and four 12-pdrs and crossed the whole of Siberia for operations against the Bolsheviks on the Kama River.*

Back home there was a hint of revolution also. In the summer of 1921 both national coal and rail strikes were dealt with harshly by Lloyd George's coalition Government who feared they might herald something much worse.

Accordingly, five fresh Royal Marines battalions were raised, the first two being composite battalions formed by drafts from Eastney. In fact, many of their numbers were made up from shipboard detachments landed from the battleships of the Atlantic Fleet. The *Warspite* was sent to Clydeside and landed her marines, but it was recorded that '...they spent most of their time playing football with the strikers...!' The cover for their assembly was the Irish situation, a precedent having been set by the formation and deployment of the 8th Battalion there from 1916. The 10th Battalion entrained for Edinburgh under Lt-Col

Above left:
An officer of the Duke of York and Albany's Maritime Regiment of Foot (The Lord High Admiral's Regiment) 1664, painted by R Wymer.

Centre:
A soldier of the Duke of York and Albany's Maritime Regiment of Foot, 1664, painted by R Wymer.

Above right:
A soldier of Prince George of Denmark's Maritime Regiment of Foot, 1685, painted by R Wymer.

Right:
A general recruiting poster c1790 by M H Bunbury.

Above:
Uniforms of the Royal Marines between 1812 and 1840 from a painting by Van Huen. In the foreground are an officer of the Regency period and a Sergeant. In the background are Royal Marines artillery and infantry officers, a trumpeter, a drummer and a private; on the right an officer in undress uniform and a gunner.

Right:
An officer 1778, Private 1779 and a Grenadier 1775 from a painting by Van Huen.

Opposite page

Top left:
Pastel portrait of General Sir John Boscawen Savage drawn c1780 when he was a Second Lieutenant.

Top right:
Bandsmen's uniforms 1825 to 1911 from a water colour by Maj J S Hicks RM.

Below:
Royal Marines and sailors returning to Portsmouth for Christmas 1805 after the Battle of Trafalgar, from a painting by Charles Stadden.

Left:
RMA officer in full dress c1874 by R Simkin.

Below:
Leaving the Officers Mess at Eastney for the Southsea Carnival in the sixties, an officer RMLI and an other rank RMA in 1906 uniforms.

Below left and right:
The original artwork for the Queen's and Regimental Colours of 43 Commando RM which were presented on 22 July 1965 and are now lodged in the Officers Mess at CTCRM, Lympstone.

Opposite page

Above left:
Uniforms c1854 from a water colour by Maj J S Hicks RM. From left: (rear) Boats Crew and Bandsman; (front) Private in guard order, Corporal in undress order and a Captain in full dress.

Above right:
RMLI officer in full dress c1864 by R Simkin.

Below:
Uniforms of 1914 by Van Huen. In foreground Major RMLI and Captain RMLI (full dress), Captain RMA and Major RMLI (frock coat). In background are a Colour Party, Colour Sergeant RMA (review order), Colonel Commandant RMLI (review order) and Lieutenant RMLI (undress order).

Above left:
A cartoon of a Royal Marines Wren (Maren) by Hynes, which is on display in the Royal Marines Museum. His cartoons appeared regularly in the Second World War magazine 'Men Only'.

Above right:
On 23 July 1964 HM The Queen reviewed the Royal Marines in the grounds of Buckingham Palace to mark the Tercentenary of the Corps. Her Majesty, accompanied by the Captain General, HRH The Duke of Edinburgh and the Commandant General, Gen Sir Malcolm Cartwright-Taylor, meets Gen Sir Campbell Hardy who, as President, commanded the Royal Marines Association contingent.

Below left:
The Massed Bands of the Royal Marines Beat Retreat on Horse Guards Parade, London, on 13 June 1973 in honour of the birthday of the Captain General. In the foreground are the personal standards of Prince Philip (left) and the Lord High Admiral (HM The Queen) (right).

This page:

Above:

The Band of the Royal Marines School of Music Beat Retreat at Broadlands, Romsey, Hampshire, when the home of the Life Colonel Commandant, Admiral of the Fleet the Earl of Mountbatten of Burma was first opened to the public on 19 May 1979, only three months before his brutal assassination by the IRA.

Below:

The last exercise of a conventional aircraft carrier with a Commando was in the Mediterranean in October/November 1983 when 40 Commando deployed in HMS Hermes.

Opposite page:

Below right:

The first occasion of a change of command ceremony, was on 11 April 1975, when the symbol of office of the Commandant General was handed over at Lympstone by Gen Sir Ian Gourlay (right) to his successor Lt Gen P J F Whiteley.

Above left:

The Royal Marines have been invited to perform Public Duties in London on three occasions, 1935, 1978 and 1986. Here a marine of 42 Commando relieves a guardsman of 1st Bn Irish Guards in June 1986. 42 Commando was the first Royal Marines unit to be armed with the new SA-80 rifle and this was the first occasion it was paraded ceremonially in London.

Below left:

A marine of 42 Commando on exercise with the newly introduced SA-80 rifle in August 1986.

Above:

The first green beret to be worn by a Royal Marine Commando belonged to Lt-Col B J D Lumsden. This beret has an officers' full dress collar badge as worn only by the officers of 41 (RM) Commando. Alongside is a Fairbairn and Sykes commando fighting knife.

The Royal Marines detachment of the battleship HMS Ramillies, *accompanied by the band, land at Milford Haven during the South Wales coal strike, which was sparked off by the government announcing their intention of decontrolling the industry. The strike lasted from 1 April to 1 July 1921.*

H E W Iremonger and remained in that area guarding naval installations against militant miners until October, when it returned to Eastney and was disbanded.

The 11th Battalion faced rather more serious problems during the so-called 'Chanak' affair the following year. The roots of the crisis can be traced back to the support given by Lloyd George to the Greek premier, M Venizelos, and his ambitions in Asia Minor, especially in the Gulf of Smyrna region. In May 1920, with the blessing of all the Allies including the Americans, Greek troops were sent to Smyrna which was taken over as a base for an ill-conceived Greek expansion into Turkish Anatolia. The Turkish indignation at such encroachment by their hated enemies led to the rise to power of Kemal Ataturk, Gallipoli's successful defender. He organised a fierce resistance, which ultimately led to the rout of the Greeks and the sacking of Smyrna. The British Mediterranean Fleet was caught up on the fringes of this fighting and the Royal Marines aboard were naturally involved. This new nationalism also threatened the Allied armies holding the neutral zone of Chanak between the two powers along the Ismid Peninsular. On 16 March 1920, 3,600 sailors and marines were landed from the fleet at Constantinople. Despite many warnings, in April the Greek army began its advance which ended in defeat a year later. They then began a build-up of troops in Thrace to which the Turks quickly responded. Royal Marines from the battleship *King George V* were landed at Smyrna on 8 September 1922 to prevent a massacre of the Greeks as the Turkish army approached, but they could only be mute witnesses to the burning of that city. The victorious Turks now marched towards the Bosphorous and the neutral zone. Panic measures were taken in London and on 11 September 1922, extra warships were rushed out and plans were made to reinforce the troops there with a battalion of Royal Marines.

The 11th Battalion arrived at Constantinople on 2 October 1922. Its infantry companies took their

MILTON ROAD
UNITED

'The Marine Who Wanted To Go To Sea'. H M Bateman's famous cartoon which for many years hung in various Corps Drafting Offices and can now be seen in the Museum.

Above left:
Marines of the Mediterranean Fleet march through Constantinople. They had first been landed on 16 March 1920 when Greece announced her intention of taking over Turkish Anatolia. Subsequently the 11th Battalion RMLI was specially raised in 1922 for service in the area in what became known as the 'Chanak Affair'.

Portsmouth Company of the 11th Battalion pass through Portsmouth Dockyard gate on their return from Constantinople on 8 September 1923. This battalion, under Lt-Col J A M A Clark RMLI, had been raised a year earlier for service in Turkey where they carried out security duties. Although the RMA and RMLI had amalgamated earlier that year, two of the drummers of the RMA Divisional Band wearing the old Broderick cap and carrying music card holders, still retain their RMA badges.

place in the fortifications known as the Ismid line protecting the city, while the RMA company was sent to the Dardenelles where they served a battery of naval guns defending the Straits. It was on these same heights that the Corps had sacrificed so many lives seven years earlier. They remained for a year following negotiations between Kemal Ataturk and General Sir Charles Harrington, but with the signing of the Treaty of Lausanne the battalion was brought home and disbanded.

The Corps, having seen another emergency through, now faced a grave crisis of their own. Ever since the Armistice and the rapid post-war run down of the fleet, the usual rumblings regarding economies began to grow. As always during times of strict retrenchment, the Royal Marines, who fitted no neat and tidy niche in the scheme of military thinking, was marked down for abolition. They had already seen their strength fall from 55,000 to 15,000 in a year and naturally it was expected that such specialists as the RM Labour Corps and RM Submarine Miners would have to go. However the baleful glare of the Treasury now turned to the Corps as a whole, and it required hard bargaining to fend off demands for its total disbandment. Eventually a reduction to 6,000 was proposed which was fiercely resisted. However the best that could be done was to accept a cut which pared the Corps down to 9,500, and this decision was promulgated in an Admiralty Fleet Order of June 1923.

It meant that one Division would have to be given up entirely. The decision was therefore taken to amalgamate the RMA and the RMLI to become simply 'The Royal Marines'. The dilapidated Forton barracks was closed and Eastney alone remained as the Portsmouth Division, while the officer lists were unified and recruiting concentrated at Deal. The abandonment of the artillery role meant that only a tiny section of officers and men would in future receive basic gunnery instruction at Fort Cumberland. The Corps would once more have to rely more heavily on the army for artillery support ashore, a retrograde step and one which rankled for several decades with former Royal Marine Artillerymen.

A committee was set up to examine future service requirements in 1924 under Admiral Sir Charles Madden, which called for the establishment of a 3,000-strong brigade able to go anywhere at short notice, and to be backed up by a fully-mobile base organisation. Such an organisation had first been proposed in April 1920 and was subsequently revived at intervals. One might have thought that the Royal Marines might here have found their natural calling, but nothing was done to implement such a far-sighted policy. Instead, their main employment continued as before the war, manning a proportion of the guns of

the fleet and carrying out policing duties afloat. Thus all the experience painfully learnt between 1914 and 1918 was discarded. This thinking continued through the next two decades of peace, and even on the outbreak of World War II, a proposer of combined operations training was told 'There will be no amphibious landings in this war!' Despite this, one of the directions to the Divisions was that they be '...especially trained to provide a striking force, drawn from either the Royal Marines Divisions or the Fleet, immediately available for use...'

Another diversion was the formation of the Royal Marine Police in 1922, using Pensioners to guard Naval Dockyards. Then with the establishment of the Fleet Air Arm of the Royal Air Force in 1924, the Royal Marines were again allowed to train as pilots. It was a Royal Marine who carried out the first ever night deck landing in a single-seater aircraft - a Flycatcher. It was made on 26 November 1929 by Royal Marine Lt Owen Cathcart-Jones aboard HMS *Courageous*. In 1934, after he had retired, he and Ken Waller broke the record for a return flight to Australia. Other Royal Marines pilots of this period were involved in operations during the Turkish crisis of 1920, actions against Chinese bandits and junks in 1926, and later in Shanghai and Palestine. Training was carried out at RAF Leuchars and until the Royal Navy got back control of its own aircraft between July 1937 and May 1939, the hierarchy was rather involved. As Major R T 'Dicky' Partridge was to record

The departure of the RMLI from Forton Barracks, Gosport, on 1 August 1923 on their amalgamation with the RMA. The Colours are being brought from the Officers Mess by QMSI B W Thatcher and received by Lt R H S Teek (King's Colour) and 2nd Lt J L R Pym (Regimental Colour). The ornate portico over the entrance to the Officers Mess (right) is now outside the Wardroom of HMS Collingwood *in Fareham, Hampshire.*

> One of the results of this awkward arrangement was that all naval flying personnel had to have an RAF commission and so we were granted a temporary one. Hence, when I started flying, I was a Lieutenant RM and a Flying Officer RAF, and later a Captain RM and Flight Lieutenant RAF.

Between the wars 46 Royal Marines became pilots, and a further 8 qualified for their wings during World War II. They served with great distinction, and no less than eleven aircraft carriers in the Royal Navy boasted Royal Marines officers as their Commander (Flying) at various times.

It was not very long after the amalgamation in 1923 that the special versatility and reliability of the Corps was called upon. Once more, as so often before, Royal Marines found themselves serving in China. This great nation, now a republic, was being torn asunder internally and the Europeans in their midst were put at grave risk. The casting aside of the

The Royal Marine School of Land Artillery was quartered at Fort Cumberland soon after World War I, although the Fort was first occupied by the RMA Headquarters on 22 October 1859. It was built by English convict labour in 1803 and the convicts then continued to live there for some years afterwards. On parade in 1921 are both field and medium guns and limbers.

Although the RMA moved into the newly built Eastney Barracks in 1867, they continued to occupy Fort Cumberland. In the 1930s the Fort was used for gunner and repository training, the art of lifting and moving heavy equipment by the use of sheer legs, gyns, tackles and ropes.

A Flight of a Fairy IIIF and two Ripons over Grand Harbour, Malta in 1931. The pilots are all RM Officers, Lts W S North, J L Moulton and N R M Skene, who were serving together in 806 Squadron in HMS Glorious.

corrupt Manchus in 1911 did not end their problems as they had hoped, and now worse things beset them. There were the continual aspirations of the Japanese, who coveted more and more of the Chinese mainland and sent increasingly large armies to nibble away at it, taking Manchuria under its wings and extending its influence along the northern coastal strip. The Nationalists arose and Soviet Russia also began spreading the net of communism into this vast population. Thus the arrival of Mikhail Borodin as Sun-Yat-Sen's adviser in 1925 soon saw an increase in the ferment. Rival armies clashed and marched

Lt M Fuller lives up to his nickname of 'Crash' as he sits on the tail of his Fairy seaplane waiting to be picked up by his mother ship HMS Courageous *in 1928. He is also reported to have jumped off the flight deck of an aircraft carrier in a bathing costume with an umbrella as a parachute for a £5 bet. He survived to retire as a Colonel in 1948.*

Royal Marines detachments were often landed from HM Ships during rioting. The marines of HMS Hawkins *are seen manning a mobile machine gun in the streets of Hankow in June 1925.*

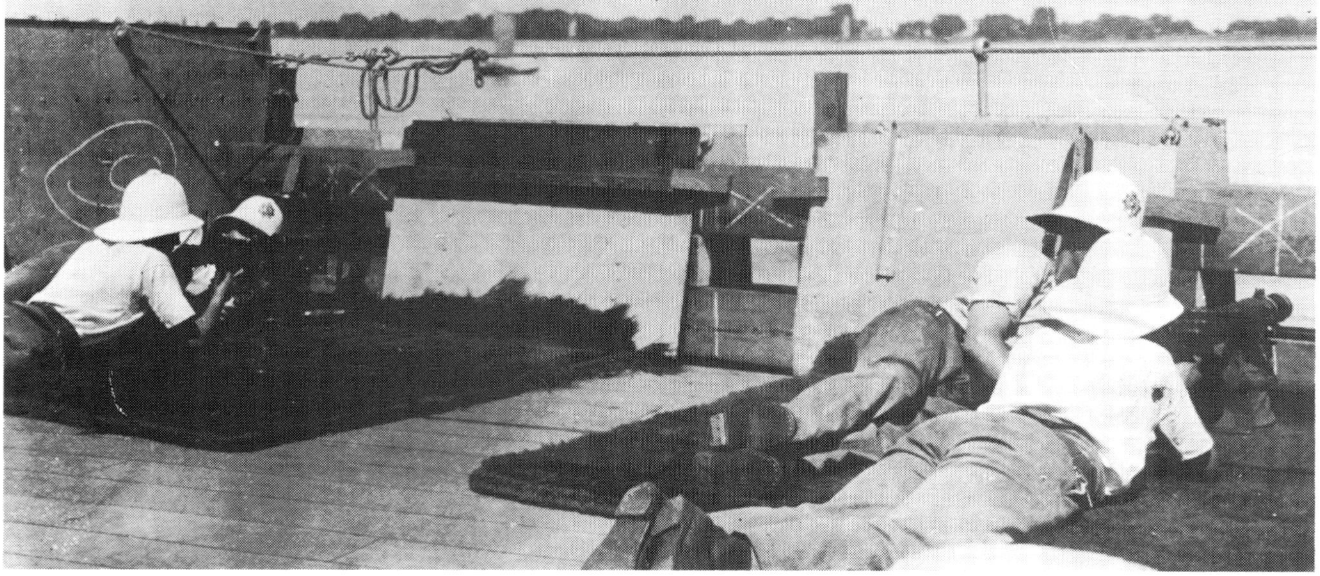

forth, and into this constant upheaval the age-old banditry and lawlessness of the war-lords continued to create local difficulties. The European traders were to be found along the two main highways from the coast, the Yangtse-Kiang and West Rivers, while on the coast they could not avoid becoming targets for all hostile forces. River gunboats were based there and warships of up to cruiser size were able to penetrate hundreds of miles up from the sea, so their Royal Marine complements were frequently put ashore hastily with rifles, bayonets and steel helmets to face rampaging mobs.

In June 1924 British and American river gunboats on the Yangtse had to act firmly to prevent bloodshed at Wanhsien where there were anti-foreign riots. A year later, following a wave of strikes and much agitation on the same lines occurring at Shanghai, an International Force was landed from warships of Great Britain, the USA, Japan and France. American pressures of their 'open door' policy led to much of this unrest and, although they always tried to distance themselves from allegations of colonialism, this did not prevent them from pushing hard where their own national interests were affected. These rumblings burst again in 1925 at Canton, Hankow and Nanking, and at Wanhsien again in September 1926. Violence once more erupted in March 1927 at Nanking, where the cruiser *Emerald* was forced to open fire with her 6-inch guns to quell rioting.

But by far the most serious incident of the decade took place at Shanghai in 1927. The advance of the Chinese Nationalist Army (the military arm of the latest nationalistic movement, the Kuomintang) against the government in Peking, again threatened the international settlements there. Angry mobs, stirred up to new bravery by the imminence of Chiang-Kai-Shek's approaching force, stormed the British concessions at Hankow and Kuikang. The C-in-C, China, requested urgently 1,000 Marines to guard them. Accordingly the 12th Battalion was raised at Eastney, taking one company from each of the three Divisions as usual, and sailed for China on 25 January 1927 to serve with the Shanghai Defence Force. This battalion was raised in a matter of days, whereas the army said they would need several weeks.

Their welcome presence at Shanghai calmed some of the fears of the Europeans, including many women and children, who were isolated while the various Chinese factions warred amongst themselves. Eventually, in April 1927, a massive strike was organised by the Chinese communist chief, Borodin, and Chiang-Kai-Shek reacted with a ruthless purge of all 'advisers' which saw many slaughtered and others expelled. Their own house put in order at Shanghai, the Nationalist advance resumed, but the Royal Marine presence was maintained for safety's sake. They remained in China for a year before returning home to disband, their job again efficiently carried out.

The Royal Marine detachment of the cruiser HMS Hawkins *man machine guns whilst on passage down the Yangtse in 1927. There had been constant anti-foreign disturbances during the preceding four years and at Nanking the cruiser* Emerald *was forced to open fire with her main armament to quell a riot.*

At home the economies demanded of the service continued to increase and a special 'Economy Committee' was set up in March 1931 which recommended cuts in service pay. That autumn this draconian policy led the Navy to face a personnel crisis when the ill-thought out and uncaringly applied pay cuts led, in some cases, to refusal of duty in the Atlantic Fleet's big ships at Invergordon and elsewhere. As in 1797, the majority of the Royal Marines remained loyal although as the cuts were applied to their meagre pay in line with the sailors they could easily have sympathised. As before and as might be expected, a few marines joined the disobedience. The label of 'mutiny' as applied by the sensational press, was too strong a word for it.

Two marines of the 12th Battalion try their hands with chopsticks whilst serving in Shanghai in 1927.

It was on Sunday 13 September, when the fleet was anchored off Invergordon, that the cuts announced by the Admiralty were revealed to the men. Many of the crews of the capital ships, *Hood*, *Nelson*, *Malaya*, *Repulse*, *Valiant* and *Warspite* plus the heavy cruiser squadron of *Dorsetshire*, *Exeter*,

Royal Marines from HMS Vindictive *hauling field kitchen equipment during field training in China in 1927.*

Norfolk, *York* and the mine-laying cruiser *Adventure* were hectored in the canteen ashore by agitators, but the day passed off without further incident. However more speech-making took place the next day. When the fleet attempted to put to sea on the morning of the 15th *Nelson*, *Rodney*, *Valiant* and *Hood* did not sail. In *Valiant* most of the Royal Marines fell in to prepare for sea, but the stokers stopped her sailing. Aboard the cruisers again all the Marines turned to at 0600 hrs but two hours later they joined the seamen and stokers in refusing to work after a 'stand-easy' or rest period. All the Royal Marines aboard the *Dorsetshire* obeyed the order, and by the evening normal working had returned to all the ships except *Adventure*, *Norfolk*, *Rodney* and *Valiant*. The next day sensational comment in the national press sparked it off again in the *Hood*. Both the Cabinet and the Admiralty dithered and left decisions to the officers on the spot, but finally orders came to disperse the ships to their home ports, and by the 17th the unrest was virtually at an end. Only the cruiser *Delhi* experienced similar problems. In the aftermath, although no communist agitation was found, as had been suggested, 121 men were disciplined of whom 24 were discharged from the service. Some of these had not taken part in the proceedings but had been labelled as trouble-makers and therefore included in the general 'clean-up' by the Board. Twenty Royal Marines from the cruiser *Norfolk*, the worst affected ship as far as the Corps was concerned, were distributed among other ships of the fleet as a precautionary measure.

Finally, it should be noted that alone among the eight ships, only Capt C A Scott of the *Valiant*, whose command was one of the worst hit, contemplated the use of force against his crew. Should the Royal Marines aboard have been ordered to fire on their shipmates, it is certain that what was really a 'two-days wonder' would have turned into something far more serious and permanent.

The self-respect of the Corps was re-enhanced in the public eye in 1935, the Silver Jubilee year of King

Royal Marines have always tried to keep their feet dry when landing from ship's boats. The detachment of HMS Suffolk *use an improvised (and unwieldy) ramp on a ship's cutter during exercises on the China coast in 1929.*

The hardiness of Royal Marines has often been commented upon in recent years, but at Chatham in 1926, a snowstorm nearly obliterated a guard of honour, parading with the Regimental Colour.

'Private Paget', the mascot of the United States Marine Corps pictured aboard HMS Cairo *during the ship's visit to Washington in 1927. This bulldog had been presented by the Corps to the USMC on 27 June that year to replace 'Sgt-Maj Jiggs' their previous mascot. 'Pte Pagett' was the fictional hero of Col W P Drury's novels about Royal Marines life published in the early part of this century.*

The C-in-C China Fleet inspects the landing party of the cruiser HMS Berwick *at Happy Valley racecourse in Hong Kong on 21 December 1933. It is interesting to note that 16 years later, after the Japanese occupation of the colony, 45 Commando made the racecourse their home for a month, the Officers and SNCOs living in the boxes! (Hospitality not horse!)*

Major N I Eastman leads the band and a Battalion RM through Gibraltar during the Fleet Review of 1930. These events ashore were a regular occurrence in peacetime, when all the ships of the Fleet were brought together for sporting and military competitions.

Capt Gene Tunney USMC (retd), former world heavyweight boxing champion, presents the United States Marine Corps Trophy on the main parade at Eastney on 1 December 1928. It was received by the Adjutant General RM, Lt-Gen L S T Halliday VC, CB and is competed for annually as the Royal Marines inter-establishment Association Football Cup.

HRH The Duke of York (later King George VI) in the uniform of a Captain RN, visits the Depot at Deal on 10 May 1928. He is accompanied by the Adjutant Capt A M Craig and in the rear is the Adjutant General RM, Lt-Gen L S T Halliday VC, CB. During his visit he watched recruits under training, lunched with the officers and then inspected the British Legion on South Green.

Royal Marines have often been depicted on postage stamps. This issue of October 1973 commemorated the 50th anniversary of the departure of the last Royal Marines from Ascension Island. The drawings are based on a series by Charles Stadden.

A church parade at Chatham in the early 1930s. Until weekly church parades were abolished after World War II, except at Deal, these Sunday morning ceremonies were watched by many of the local population at the Divisions, and were an established feature of the visitor scene. This parade at Chatham shows the band trooping, at this time without drums in front, but with buglers in rear. The barracks guard have turned out wearing blues and musketry order.

The detachment of the battleship HMS Barham *parading in marching order on the quarterdeck prior to landing at Haifa in Palestine in July 1929. Each year a Royal Marine striking force of two battalions formed by ship's detachments of the Mediterranean Fleet exercised ashore.*

George V, their Colonel-in-Chief, when the Royal Marines were invited, for the first time, to perform ceremonial London Duties. The Corps was asked to provide a London Battalion from all three Divisions to replace a Guards battalion who were to be away on manoeuvres. Although the King and Queen were absent from the capital during this period the honour afforded the Royal Marines of mounting guard every third day at Buckingham Palace, St James's Palace, and providing picquets at the Bank of England and the Hyde Park magazine was deeply appreciated by the Corps as a whole. They also exercised a jealously preserved right, granted to them in their earliest days, that of marching through the City of London with bayonets fixed, drums beating and Colours flying. In consolidating their link with the premier city of the Empire, the sad death of the monarch, a true friend of the Corps, a year later, also saw the Royal Marines involved with the ceremonials of the funeral, and subsequently the Coronation of King George VI in 1937.

Meanwhile sterner issues were facing the country in the Mediterranean. The arrogant boasting and towering ambitions of the Fascist clown Mussolini, dictator of Italy, had led to tensions during the summer of 1935. His threatened invasion of Abyssinia and the pathetic posturings of both the League of Nations and the British and French governments in trying to thwart the move, had led for a time to a threat of war. The British Mediterranean Fleet was in prime condition for such a combat and certainly did not fear it. Eventually, however, lack of similar resolve at home led to weak and ineffective sanctions, while the Italian conquest proceeded without much hindrance from anyone. The fleet left its base at Malta, where it was too vulnerable, and sailed for a waiting position at Alexandria. However the facilities available there for a proper fleet base caused concern and the much neglected Mobile Naval Base Defence Organisation still did not exist.

The provision of the MNBDO had been re-examined in 1929 and had been made a firm commitment for the Royal Marines at that time. It was hoped that within 48 hours of the fleet's arrival at a selected harbour, the MNBDO would be set up and running. An MNBDO manual had indeed been called for but it had not been written in 1935. In the light of events that actually took place, the concept had to be revised and the manual did not appear until 1936. As Admiral Fisher commented, 'the peacetime organisation and training of the MNBDO has been based on three major assumptions, all of which turned out to be wrong.' The need for its own anti-aircraft defence was one item to be added to this (as yet academic) scheme. The weapons, personnel and equipment for such an enterprise had been agreed in 1931. The basic requirements were specified as specialised landing craft, six coastal defence guns, 22 anti-aircraft guns, 33 searchlights, provision for the building of 100ft long piers, construction material for roads, anti-torpedo baffles and anti-torpedo booms and nets, controlled minelaying facilities, a signals section and a headquarters staff. To transport all this equipment, two specially converted merchantmen, *Bencruachan* and *Bellerophon*, were prepared.

The Massed Bands of the Mediterranean Fleet Beat Retreat in Palace Square, Malta in 1934. It was at a similar ceremony in Gibraltar that year that Fleet Bandmaster A C 'Jimmy' Green's arrangement of the Sunset Call for bugles and band was performed for the first time. This is still played regularly at Royal Marines functions.

A composite Royal Marines Divisional Band playing at St James's Palace when the Royal Marines carried out Public Duties in London for the first time in their history in August / September 1935, Jubilee year. Here they can be seen relieving the 2nd Battalion Coldstream Guards.

In the panic caused by the 1935 Abyssinia crisis, a substitute organisation was hurriedly cobbled together. Under the command of Brig H L Tripp DSO,

HRH The Prince of Wales, in the uniform of the Seaforth Highlanders of which he as Colonel-in-Chief, presents the King's Badge to Rect A J Payton of 212 Squad at Deal on 5 July 1935.

The only visit King Edward VIII paid to the Corps during his short reign was to Eastney in 1936. From right to left: HM The King, Col Comdt Alan Bourne (Commanding Portsmouth Division), Commander Lord Louis Mountbatten, Lt-Gen Sir Richard Foster (Adjutant General RM), Admiral of the Fleet Lord Chatfield and Admiral of the Fleet Lord Keyes of Zeebrugge fame, (Hon Col Comdt Portsmouth Division).

MC, this organisation named The Base Defences Mediterranean, was a composite force from all three services. The Royal Marines part in this force was the provision of the commander, 1,600 officers and men to man the 6-inch heavy coastal guns and the light artillery defences, the AA artillery and accompanying searchlights as well as signals and a landing and transport company. On landing at Alexandria, only the coastal guns were actually emplaced, the searchlights and AA units being retained on board for possible use at the proposed forward base, Force 'X'. This was to be at Navarino Bay on the west coast of the Peloponnese (and not Suda Bay on the island of Crete as it was to be six years later) which belonged to Greece, and thus required the blessing of their government before it could be put into a similar state of defence. As the Greeks were scared of the Italians, this base never materialised and eventually all units were landed at Alexandria as originally planned.

By July 1936 the crisis had passed and the Base Defences were dismantled, the Marines returning home. Although there had been little action, the need for such an organisation was now clearly recognised and could no longer be shelved. Not much was done

The early days of landing craft in Langstone Harbour. BMB No 1 unloads a Morris-Martell tank directly onto the beach in 1929. Only eight of this version were ever built for the Experimental Mechanized Force in 1926/27.

immediately but the groundwork was laid. The development of the first custom-built landing craft (which the Japanese had perfected and been using for years in China) was undertaken at Fort Cumberland, and pier building equipment was studied. The concept of the Mobile Naval Base Defence Organisation was taken out of pigeon-holes yet again and dusted down.

As crisis followed crisis during the next few years (the German demands on Czechoslovakia and the shameful Munich agreement, the Spanish Civil war, the rape of Albania and Japan's continued aggression against China) it became more and more clear that war was inevitable. Belated re-armament was made in an effort to repair twenty years of total neglect in twenty months. It failed to beat Hitler's timetable, but at least some positive results flowed from it. As far as the Royal Marines were concerned, the 1st RM AA battery was sent out to Alexandria (minus guns!) while at home a RM Special Gunmounting Party began to site guns at strategic points on the east coast of England. The bulk of the marines work continued as before, and it is clear in hindsight that the opportunity to play a prominent part in the development of landing-craft had been squandered. A Landing Craft Committee had been set up in 1924, and the Inter-Services Training and Development Centre, under Capt L E H Maund RN in 1938, but by the outbreak of war the following year, only two landing exercises had been carried out and both of these involved the army. Only prototype LCAs and LCMs had been developed.

As the sands ran out in the autumn of 1939, mobilisation orders once again sped to the far corners of the land. By 3 September the Corps was able to muster 12,390 officers and men, with a further 1,082 from the Royal Fleet Reserve and 2,406 Royal Marines pensioners. Conscription was obviously urgently required if the Corps was to meet even its limited basic obligations and a new camp was established at Exton (later renamed Lympstone) in Devon for that purpose, the first trainees arriving on 23 February 1940. However, the 'rehearsals' of 1935 and 1938 had stood the Navy in good stead, and if the Royal Marines were somewhat under-strength most of the men were at least in the right place at the right time when the signal 'Total Germany' was flashed from Whitehall to the fleet on 3 September 1939.

Beach landings were a regular part of the training from ship's detachments. In the mid-1930s Royal Marines, wearing 1908 pattern web equipment over their Khaki Service Dress and long puttees, and armed with the SMLE, make a wet landing from a ship's cutter.

Amphibious operations always have the problem of dealing with soft beaches. During an exercise at Portland in 1927 this Hathi Tractor found the going poor. These heavy tractors of the RM MNBDO Experimental Unit had an 11 litre, 6 cylinder engine and cost £3,000 each.

Royal Marines have always played a major part in the annual Royal Tournament in London. In 1934 they put on a display of the drill of 1664, 1805 and 1934. Capt (later Maj-Gen) H T Tollemache is the mounted officer while Capt (later Col) F B Pym watches from the left, all suitably attired in 1664 period uniforms.

Opposite page below:
Royal Marines regularly took part in Naval Field Gun competitions during the 1920s and 1930s and here they are seen at Whale Island competing for the Brickwood's Trophy. Royal Marines teams competed at the Royal Tournament until 1934, when they were withdrawn on manpower grounds in favour of Stoker teams.

Hathi (the Urdu *for elephant) Tractors tow heavy gun barrels from Fort Cumberland to Eastney in 1929. The tractors had an unladen weight of 5 tons and could tow up to 40 tons and they were used for dragging gun mountings into position.*

Right:
Royal Marines have been portrayed in many films through the years including 'Girls At Sea' (from the stage success 'The Middle Watch'), 'Cockleshell Heroes', 'The Man Who Never Was' and 'Piccadilly Incident', but the Crazy Gang appeared as Marines in the film 'Alf's Button Afloat' in 1938. From left (top) Bud Flanagan, Chesney Allen, Teddy Knox. (Bottom) Jimmy Gold, Jimmy Nervo and Charlie Naughton.

133
100

CHAPTER 9

1939-1945
VINDICATION

'Meet a representative of the greatest Corps in the world!'.

Winston Churchill to Marshal Stalin at Yalta.
8 February 1945

One ruling that affected the expanding Corps during the first 18 months of Word War II was that none of the 'Hostilities Only' men, who began to join in October 1939 and reach operational units six months later, were to be drafted for sea service. This ruling was introduced for the best of motives so as not to dilute the front line working of the Corps but, in fact, it was to have an adverse effect. The traditional heavy demands for the fleet made massive and ever-increasing inroads into the established complements, while casualties and losses could only be replaced by regular volunteers. This in turn meant that the supply of time-served experienced men, so necessary to instill training and traditions into both new volunteers and HOs alike, was seriously reduced. This self-inflicted ham-stringing of the Corps was only remedied when the first HOs went to sea in January 1941, while the first HO Officers did not go until June 1943.

Secondly, the lost opportunities to establish a proper amphibious role left the bulk of the land-based units 'in the air' for a long period, when it might have been thought that Britain's desperate plight cried out for such a disciplined and well-trained body of fighting men. Although the numbers of the land-based Royal Marines actually exceeded those serving during the trench warfare of 1914-18, their actual employment in war zones was far less, and it was with the big ships of the fleet that the Corps saw the bulk of their action between 1939 and 1943.

Initially it seemed as if this new war would be a re-run of the Great War, albeit on a smaller scale. The fleet was smaller, the armies were smaller and there was the comparatively new and still unknown element of air power. Early tragic losses of major warships followed the 1914 pattern with heavy Royal Marine casualties being suffered with the loss of the battleship *Royal Oak* at Scapa Flow and the aircraft carriers *Courageous* and *Glorious* at sea during the first six months. Equally, the only surface ship success came from South America, the Falklands victory of 1914 being mirrored in the chase and destruction of the *Admiral Graf Spee* in 1939. In this action, the marine-manned turrets aboard both British cruisers, the *Ajax* and *Exeter*, were hit with heavy casualties as a result.

The officers of LCG(L) 1007 brief the ship's company before the craft started the channel crossing on 5 June 1944 to the Normandy beaches. Over 10,000 Royal Marines took part in Operation 'Overlord' with more than two thirds of the minor landing craft being manned by the Corps.

The work of the Special Gun Mounting Party, renamed the Royal Marine Fortress Unit, continued apace and the Royal Marine Engineers were re-formed to back this up, providing specialist skills at undeveloped naval bases like Scapa Flow. Another merchant ship, the *SS Mashobra*, initially based in the Orkneys, was requisitioned to provide this force with mobility. The gun mountings, finally demolished after twenty years work in 1938, were replaced. The go-ahead was at last given for the formation of a Royal Marines Brigade, mooted long ago, while the anti-aircraft functions of the Corps were concentrated at home. As there were still no guns for the 1st AA Battery, they were brought home from Egypt and split up among newly-forming batteries in the UK. A light AA battery, the 21st, was formed and equipped once more with the old 2-pounder pom-poms (whose

Military training was often carried out on the parade ground. Major C F Jerram, who later commanded the Officers Wing of the RM Military School at Thurlestone, instructs a company commander's course. Regular SNCOs who were to be commissioned are learning the use of the prismatic compass at Deal early in 1940. Note the slung respirator carried at all times. In the centre is Sgt (later Capt) Tommy Lenham, a well known Corps boxer who became an international amateur boxing referee.

Marines of Kent Battery of the 1st RM Coast Defence Regiment mount 6" guns on the seafront at Lowestoft, Suffolk in 1940. Such batteries were set up along many parts of the east coast of England as a defence against a possible German invasion. Kent Battery, along with Devon, X, Y & Z batteries later saw service in such areas as Egypt, Crete, Adu Atoll and Ceylon as part of the Mobile Naval Base Defence Organisation (MNBDO 1).

chief attributes seemed to be the need to hit them with a hammer after every third round to keep them firing!) They were based on the Tyne to deter German bombers from attacking the ship-building yards that, in those days, lined its grubby waters. The corsair activities of the German pocket battleships, disguised merchant raiders and long and short-range bombers, led to the establishment of Defensively Equipped Merchant Vessels (DEMS) which saw RM pensioners joining those of the Royal Navy, going to sea to man weapons hastily bolted aboard our merchant navy vessels.

In January 1940 that hoary old chestnut, the MNBDO was again put forward. Indeed two such were now contemplated. The proposal was for a complement of 202 officers and 4,089 men and plans were brought forward to form the first of these. The first 2,000 men joined in February 1940 and were based at Fort Cumberland and empty holiday camps on Hayling Island. However no great priority was allocated to it even now, a decision which was yet again to prove a costly mistake when Hitler launched his surprise attack on Norway on 9 April and a base had to be hastily set up by the British at Harstadt.

Troops had already been embarked in a cruiser squadron in readiness to carry out 'Plan R4', the occupation of the Norwegian ports of Bergen, Stavanger and Trondheim. On the sighting of what turned out to be the German invasion fleet, Winston Churchill, the First Lord of the Admiralty, on his own initiative ordered that these troops were to be disembarked so the cruisers could join the fleet. This premature order caused considerable embarrassment soon afterwards, when the cruisers arrived off Norway and found the Germans well established ashore. As usual, the Royal Marines filled the breach and they were part of 350-strong force, including a 3.7-inch howitzer battery, under the overall command of Capt W F Edds RM, put ashire at Namsos from the cruisers *Glasgow* and *Sheffield* until relieved by army units a few days later. Capt G W Wilson, in charge of the howitzer battery, kept the evacuation destroyers waiting with the words 'It is not the policy of the Corps to leave its equipment in enemy hands.' And nor did they!

The battleship *Resolution* also landed her Royal Marines detachment, commanded by Major G V Walton at Aa Fjord and they operated on the Narvik

The only occasion on which a reigning monarch has presented the King's Badge personally was on 19 December 1939 at Eastney. King George VI pins the badge, awarded to the best recruit, to Rect F E Woods of 353 Squad. A marine continues to wear this badge for the rest of his service even if subsequently commissioned.

front most successfully, rescuing six British seamen from two captured trawlers who were in a German aircraft that made a forced landing. They tracked down the German party and their POWs to a Norwegian hamlet, surrounded and stormed it from front and rear, to capture the enemy garrison and set the British trawlermen free.

One of the most striking contributions to this campaign was made by Royal Marine aviators serving with the Fleet Air Arm. Capt R T Partridge flew a Skua dive bomber over Norway and participated in the sinking of the *Königsberg* along with fellow Royal Marine pilots Capt E D McIver and Lt L A Harris; he

Squads of Royal Marines at drill at Chatham in 1940. Note the variations of dress with both khaki service dress and battledress being worn. The 'BM' on the roof denoted the Brigade Major's house and was for the benefit of fire-watchers and not enemy aircraft!

Royal Marines from HMS Resolution *landing from the trawler* Northern Gem *to rescue British Prisoners of War in Norway in 1940.*

The volunteer drum and bugle band of the 8th Battalion RM in 1942 with front row from left, the Adjutant (Capt J A Taplin), the CO (Lt-Col S G B Paine), the Drum Major (Cpl L G Rayfield) and RSM R Keeble.

also shot down a Heinkel He111 bomber, being shot down in return. After landing safely he managed to take the entire Heinkel crew prisoner for which he earned the DSO. Partridge escaped to join his squadron at sea again, only to be shot down once more during the abortive bombing of the battle-cruiser *Scharnhorst* on 10 June. He spent the rest of the war in a prisoner of war camp along with Lt Guy Griffiths and former Royal Marine Lt H M A 'Wings' Day (subsequently Wing Commander) who had been awarded the Albert Medal for the rescue of two wounded men below deck after *HMS Britannia* was torpedoed off Gibraltar on 9 November 1918.

On 17 April a makeshift battalion of 700 sailors and marines from the refitting capital ships *Barham*, *Hood* and *Nelson* was given the rather inappropriate codename 'Primrose'. Commanded by Lt-Col H W Simpson, they were hastily embarked and landed at the small Norwegian ports of Alesund and Andalsnes, where they established AA and searchlight positions only to be evacuated on 29 April. The Royal Marine Fortress Unit, commanded by Lt-Col H R Lambert was despatched post-haste to Skaanland and began establishing base facilities, but their ship was sunk by air attack. Lambert was wounded and the unit withdrawn on 26 May, its work incomplete. They returned to form the cadre of the second proposed MNBDO back in England.

Marines pull through 'Winnie', the gun named after Winston Churchill, then Prime Minister. Soon after the fall of France in 1940 the RM Siege Regiment, under Lt-Col (later Maj-Gen) H D Fellowes, was formed to man the static 14" naval guns, nicknamed 'Winnie' and 'Pooh' at St Margaret's Bay near Dover in Kent. Their maximum range of 27 miles enabled them to shell the French coast but they could also be used in an anti-invasion role. Later three mobile 13.5" guns 'Sceneshifter', 'Peacemaker' and 'Gladiator' were installed on the railways in East Kent.

Re-establishing their old prerogative of 'first in - last out' the final British troops to leave Norway when the campaign came to a hurried end in June were two platoons under Lieut E D Stroud, which acted as a rearguard at Verma. Although the bulk of these got away without incident, the early arrival of the German advance guard found Stroud, a corporal and three marines with two Lewis guns and three rifles between them, forced to hold a pass. They had to make a stand for three and a half hours to keep the strong enemy force at bay.

Denmark and the strategic dependencies of the Faroes had been taken over by the Germans, which meant that Iceland had to be occupied by Marines. The newly-formed 2nd Battalion was sent, reinforced by men from the fleet and others hastily enlisted from training camps in southern England, together with Royal Marines coast artillery and AA units under the name of Force 'Sturges' and they landed at Reykjavik on 10 May, a total of 40 officers and 775 men. They were relieved by the army on 19 May, but the RM Fortress Unit were sent to Iceland on 12 July, preparing gun sites and installing eighteen 3-inch to 6-inch guns, before returning to UK in October 1940, then being absorbed into MNBDO 2.

With potent reminders of their intervention at Ostend in 1914, hastily prepared parties of Royal Marines were carried to the Hook of Holland at the start of the London Blitz to cover the evacuation of the Dutch Royal Family, their government and the Dutch bullion reserves. Similar groups covered the Belgian and French withdrawals with marines providing port parties and rearguards at Boulogne and Calais, taking heavy losses.

After the humiliatingly quick expulsion of her army from Europe, followed by the surrender of France, Britain was faced with the first serious threat of invasions since 1802. Among the many decisions taken about her defences was the immediate establishment of coast batteries to cover each and every possible landing place, be it harbour, bay, cove or inlet, along the southern and eastern coastline, from Cape Wrath round to Land's End. Nobody had ever conceived such a need pre-war and there was an acute shortage of the necessary heavy and medium guns to implement such a policy. The Army Ordnance Department had virtually nothing to offer at this stage of the war, but fortunately the Admiralty had been more prudent and fore-sighted. When the great fleet of 1918 had been sent uncaringly to the breakers by the politicians, Their Lordships, in their wisdom, had ensured than many of the great guns from those ships were carefully greased, sealed and stored away. Their good husbandry was now an undeserved bonus for the government and as fast as these old 9.2-inch, 6-inch, 5.5-inch, 4.7-inch and 4-inch pieces could be extracted, they were emplaced. The RM Fortress Unit dug in some 133 guns during the dark days of 1940. The next problem was the manpower required for so many batteries and it was the Royal Marines who were initially called in to assist until the army coastal artillery could recruit and train enough of their own gunners.

The Germans quickly established their own heavy guns along the French coast from Cap Gris Nez to Calais. Soon they had three 16-inch, four 15-inch and eight 11-inch guns in place from four batteries, including a railway battery. These monster guns began to menace the coastal convoys and to shell Dover itself. Already, in May 1940, the Premier had called for long-range guns to be mounted at St Margaret's Bay, near Dover, to command the straits. This was based on a plan of 1918 to mount two giant 18-inch guns there, which was never implemented. The British response again fell on the Royal Marines. Among the old pieces found suitable for such work were four 14-inch breech loading Mk VII barrels (presumably spares for the battleship *Canada*, an ex-Chilean ship taken over and the only Royal Navy battleship to carry this calibre gun in World War I.) However only two mountings were found for these guns.

On 24 June 1940 work began to establish two of these weapons on St Margaret's Golf Course. Observation posts, fire control buildings, power houses, communications, shelters and living accommodation all had to be constructed from scratch, and a special railway spur also had to be built and bridges strengthened on the main lines. The guns were shipped down to Sittingbourne on the gun barges *Gog* and *Magog* and then sent by rail to the site where three heavy lifting cranes were used. The mounting for the first arrived on 10 July and the piece was emplaced by 7 August. Each mounting was in two 75-ton sections with 15-ton cradles while the piece itself weighed 97 tons. It was housed in a steel barbette which had a light pitched roof. Nicknamed 'Winnie', after Churchill who had taken a paternal interest in this battery, it carried out its first shoot

The Royal Marines Band playing in the waist of HMS Hood, *one of their less spectacular, though essential duties aboard a battlecruiser of the 1930s and 40s. The ship was sunk in action on 24 May 1941 by the German battleship* Bismark *with the loss of 1,417 lives including 134 Royal Marines and the whole of her 17-strong band. At any point during World War II up to 1,000 musicians in the 63 bands were serving afloat and the Royal Marines Band Service suffered the greatest percentage casualties of any other individual arm of the Services, a total of 225 killed or died of wounds.*

on 22 August. The second piece followed towards the end of 1940 and began firing on 19 February 1941. The 14-inch shells were loaded with a derrick which lifted them from small railway trolleys. The guns were super-charged with extra cordite to give an absolute range of 48,000 yards, just over 27 miles enabling them to hit the French coast. Inevitably the second gun was nicknamed 'Pooh'. It was in an open mounting but, despite counter bombing and shelling, neither gun was hit although several casualties, both dead and wounded were suffered over the years. On 9 November 1942 the enemy scored direct hits on 'Pooh's' ready magazine and 'Winnie's' power house.

Once in position as 'A' Battery, the 14-inch guns were served by the Royal Marine Siege Regiment commanded by Lt-Col H D Fellowes. By August 1942 they had been reinforced by two 15-inch pieces and in July 1943, the Marine guns went on a 'Care and Maintenance' basis.

In 1918 the army in France had used super-heavy railway guns built by Vickers-Armstrong. The pieces were 14-inch Mk III and Mk IVs of which three 'Gladiator' 'Peacemaker' and 'Sceneshifter', had been brought back to England and stored. In November 1939 Col Montague Cleeve discovered two of them, minus their barrels, in a long-disused transport shed

A passing out parade at the new RM Reserve Depot, Lympstone in 1940. These Hostilities Only recruits are at the first movement of the 'Present Arms' while Band Sergeant Wally Smith (left) is waiting to play for the General Salute. The Band was made of pensioners (average age 46) from the Divisions, recalled for wartime service.

at Chilwell, Nottingham, along with seventeen 9.2-inch guns. It was thought that 13.5-inch pieces could be mounted on them. These two were also assigned to the RM Siege Regiment and detachments of 'B' Troop, under Capt N D Charrington, attended the School of Heavy Artillery at Catterick in August 1940.

The first of the 13.5-inch guns from Elswick Ordnance Works was mounted on 8 August and 'Sceneshifter' was commissioned on 20 September. The full gun train consisted of the gun, ammunition wagon, cordite wagon, fitting van, galley, dining coach, and brake. The sites for the new battery were the Guston and Lydden tunnels on the rail links between Deal and Dover, with the Stone Hall sidings used as a maintenance and repair gun park at Cliffe, near St Margaret's, and another at Swingate. The two looped firing spurs were ready for operations by September 1940, No 3 Troop manned 'Sceneshifter' and No 4

Unloading stores from the Clan Forbes *by LCAs at Adu Atoll in the Maldive Islands in 1942. 1st RM Coast Regiment and the Landing and Maintenance Unit of MNBDO 1 arrived on the tropical island in September 1941, the latter being joined by Royal Marine Engineers in the autumn of 1942. They built Gan Aerodrome, abandoned after World War II, but refurbished in the 1950s as an RAF staging post.*

Troop 'Peacemaker', which arrived on site on 27 November. During an early practice shoot on 10 December, counter-battery fire from across the channel scored a near miss which damaged the latter. Plans were made that should the invasion come the guns would remain in the forward zone for as long as possible, taking cover in the tunnels during dive-bombing attacks. They were then to fall back to defensive positions at Bekensbourne and Ashford, from which they could cover Dover Castle, Sandown Castle and Walmer among other prime objectives.

By the end of February 1941 they had actually moved back to these positions, while 'Gladiator' was finally commissioned by 5 Troop at Lydden on 8 May 1941. A mass shoot by all five 'super-heavies' against German invasion barges at Calais on 10 May failed, owing to problems with the spotting aircraft. In July a new Corps Commander Coast Artillery, army Brig W Swinnerton arrived, and the guns were again resited. The Royal Marines connection with these guns came to an end with the handover to the Royal Artillery on 17 November 1943.

While the threat of invasion hung over the country, other Royal Marine units went to their defensive battle stations along the south coast. This included the 11th Battalion, which had originally been assigned to the role of Land Defence Force for MNBDO 1. The cadre of this battalion formed at Fort Cumberland in January 1940 before proceeding first to Gosport, then to Plymouth, thence to Hayling Island. In June it took up front-line duties during the Battle of Britain, when it held a shallow coastal section stretching from Sandwich to Kingsdown on the Kent coast. Here the battalion remained until January 1941, under a bewildering succession of commanding officers. The 1st RM AA Regiment also saw service in the Dover area, as did the 11th Searchlight Regiment based on Deal.

Another abortive expedition, as fruitless as Norway had been, was mounted against the Vichy French at Dakar on the West African coast in September 1940. The 101st and 102nd Royal Marine Brigades were assigned to help General de Gaulle and his Free French force, who hoped to raise the African colonies against Marshal Petain. Royal Marines gunners also manned the guns of the battleships *Barham* and *Resolution* which carried out bombardments in support. It turned into another farcical failure and yet another humiliating withdrawal: the Royal Marines never even got ashore. The 2nd and 3rd RM Battalions remained in

Freetown until January 1941 in case operations were required in the Cape Verde Islands.

Earlier in the year, in May 1940 the RM Brigade had been split into two to become the 101st and 102nd RM Brigades. Now a new brigade, the 103rd, was to be raised at Exton, the whole force forming the Royal Marine Division, whose first commander was Major-General R G Sturges. 1st and 5th Battalions formed 101 Brigade, 2nd and 3rd Battalions the 102nd, all being concentrated in Scotland, while 7, 8 and 9th Battalions (at Dalditch) and 10th Battalion (in Glasgow) had formed 103 Brigade by April 1941.

For most Royal Marines, however, the period of 1940-41 meant sea service, manning the guns of the battleships, battle-cruisers, aircraft carriers and cruisers of the fleet. Marine gunners saw action in the many surface engagements, taking more than their fair share of casualties in these fights, particularly in the heavy attrition caused by German dive-bombers against which the fleet proved highly vulnerable. Royal Marines gun crews thus served at the sinking of the *Bismark* in May 1941, and when the mighty *Hood* blew up after a single hit every one of her Royal Marines detachment of 164 died with her. Likewise, many were killed when the *Prince of Wales* was hit. Another ship which suffered heavily was the heavy cruiser *Berwick*. When she was engaged against the Italian battle fleet at Spartivento in November 1940 her Marine turret was hit, the crew being burnt to death. On her way home she was caught in another

In February 1942, Major Duncan Johnston and 100 volunteers from the 1st RM Coast Regiment of MNBDO 1 formed Force Viper *'for special service of a hazardous nature'. They took over a number of motor launches for patrols and operations up the Burma rivers and were the last to leave before the fall of Rangoon. Here Indian troops are embarking near Kalewa after the battle of Shwegyin on the Sittang river. The force scuttled their launches before withdrawing on 17 May and marched out across the mountains reaching Calcutta on 25 May, 58-strong.*

On 26 June 1940 Commandant HRH The Duchess of Kent inspected the Royal Marines Wrens, known as 'Marens' at RM Barracks Eastney. The first Wrens had served with the Corps during World War I. At first their wartime uniforms were far from flattering and those attached to the Corps wore the 'Globe & Laurel' badge on their floppy hats or berets.

In the picture below, by now, every bit as smart as their male counterparts, Royal Marines Wrens march through Portsmouth during 'Wings for Victory' week in 1943. They wear the 'Globe & Laurel' badge on a red background in place of a ship's cap tally.

A passing out parade at Dalditch Camp, between Exmouth and Budleigh Salterton, Devon, after the wartime 14-weeks course. The first Squads X 111 and X 112 passed for duty on 1 May 1941. At various times this wartime nissen-hutted camp housed Nos 7, 8 and 10 Battalions, a company of RM Engineers, 31 & 32 Light Batteries RM, in addition to elements of 101, 102 and 103 Brigades who took up temporary residence. By 1943 20 and 22 RM Training Battalions were based there and the camp finally closed at the end of 1946.

A Marine's kit is an elaborate affair even in war-time. Kit inspections were held regularly, sometimes soon after 'Both Watches of the Hands' at 0600 on the quarterdeck. This marine in 1940 shows what a perfect layout should look like. He wears one good conduct badge for three years service at that time, the old 3rd class gunnery rate (known as a sauce bottle) and a marksman's badge for small arms. The blue side cap was re-introduced during the war to replace one blue peaked cap at sea.

fierce duel with the German *Admiral Hipper* on Boxing Day, when her turret was again hit by 8-inch fire with similar results. More casualties were suffered during the bombing of the *Illustrious* and *Gloucester*, and the sinking of the *Southampton* by German Stuka dive-bombers in January 1941, and also in the many ships of Admiral Cunningham's Mediterranean Fleet when it was attacked from the air off Crete in May. Others died when the *Barham* was torpedoed and blew up in the Gulf of Sirte later that year. However, they were able to take their revenge when manning the guns of the fleet that defeated the Italians at the battles of Calabria, Cape Matapan, Sirte and several other skirmishes the whole length of that much disputed sea.

Indeed, once the threat of invasion had lifted the whole focus of the war shifted to the Mediterranean. MNBDO 1 arrived in Egypt, via Durban, in April 1941. Six weeks later, after some tactical loading of their ships, they moved to Suda Bay in Crete to set up a forward base. But due to a lack of shipping only half the force, about 2,000 Royal Marines, had reached the island when the German airborne invasion started. Under Maj-Gen E C Weston these coast defences and AA gunners, searchlight units and signallers put up a very spirited resistance, often fighting with 'butt and bayonet' as infantrymen. General Weston assumed command of all land forces in Crete during the final few days. A hastily-assembled composite battalion, led by Major Ralph Garrett, fought a rearguard action during the evacuation, and half of those who survived the fighting had to be left behind to be taken prisoner when the Commonwealth defences were finally overwhelmed. Garrett himself managed to escape in an abandoned landing craft to which a makeshift sail of blankets was jury-rigged. With 137 companions he sailed the 250 miles to safety, with men going over the side in relays to keep her on course.

What remained of MNBDO 1 had no role until refitted, so the 1st Coast Regiment and the Landing and Maintenance Company was shipped east under Col W B F Lukis and Lt-Col L O Jones. They carried out Operation '*Shortcut*' to set up bases at Addu Atoll, Diego Garcia and the Seychelles in the Indian Ocean during the Spring of 1942. Meanwhile, during the

The Royal Marines detachment of HMS Illustrious *preparing to land in Madagascar following the occupation by the British Forces. The army operation to secure the island on 6 May 1942 against possible Japanese occupation was halted by the Vichy French defenders, and the timely arrival of Royal Marines landed from HMS* Ramillies, *under Capt Martin Price and Lt Jimmy Powell, to 'cause maximum noise and confusion' paid off.*

Marines of 40 (RM) Commando on their way to the abortive raid on Dieppe on 19 August 1942. This was the first Royal Marine Commando operation as part of a larger raiding force. Few of the Commando actually landed, although a RM provost party under Lt-Col R G Parks-Smith, a pioneer of parachute landing, got ashore to maintain discipline and order over all personnel.

latter part of 1941, 11th Battalion under Lt-Col E H M Unwin had been left in Egypt kicking its heels impatiently, and training repeatedly and hard as an amphibious striking force in true Marine style. It practised on the Great Bitter Lakes until being transferred to the Lebanon early in 1942. It carried out a small-scale raid on Kuphonisi Island on 15/16 April, but was sacrificed in an ill-conceived and badly executed attack on Tobruk on 13/15 September. Due to inadequate landing craft and a misunderstanding in the landing orders, most of the first wave foundered on the beaches. Those that got ashore fought gallantly until overwhelmed, Major Jack Hedley, a renowned Bisley pistol shot killed five Italians with six shots in taking a strongly-held defensive post. Most of the second wave were lost when the destroyers *Sikh* and *Zulu* carrying them were both sunk. This senseless waste of a battalion of nearly 400 highly-trained men did nothing to improve Royal Marines morale, already low after months of enforced idleness. Over 220 officers and men, including the Commanding Officer were lost or taken prisoner at Tobruk.

When Japan entered the war on 7 December 1941, a whole new tide of disasters rolled over us with the loss of Hong Kong, where small numbers of Royal Marines fought as infantry, Borneo, Burma, Malaya and Singapore. When the capital ships *Prince of Wales* and *Repulse* were sent to the bottom of the South China Sea by Japanese naval torpedo bombers, the casualties included 67 Royal Marines. The survivors from the two ships were re-organised ashore and fought as two infantry companies commanded by Royal Marines Capts R G S Lang and C D L Aylwin. Lt R J L Davis from the *Repulse*, commanding a Royal Marines composite platoon, joined up with a small force under Major Rose of the 2nd Battalion, Argyll and Sutherland Highlanders, in river patrolling on the west coast of Malaya on Christmas Eve. The remainder joined up with other remnants in early January to form a composite battalion, which was inevitably dubbed the 'Plymouth Argylls' after the football team, both ships being manned from Plymouth. They took part in the last stand in Singapore before being taken into a cruel captivity from which many never emerged.

In the wake of the disaster in Burma, another improvised group of 107 from MNBDO 1's Coast Regiment was formed to man armed river launches including some stern wheel paddle steamers on the Irrawaddy river under the command of Major D Johnson. During March 1942, covering the army's retreat, they fought a war of surprise and ambush over the same ground as the victorious campaign of 1885. Known as *'Force Viper'*, they did score a small victory over the advancing Japanese at Henzada on 17 March, but in turn were badly mauled at Padaung ten days later, suffering heavy losses. By 17 May they had reached Sittaung, 350 miles north, and could retreat no further. They burnt their boats and marched across the jungle-clad mountains to India, the remnants, 48 strong, finally reaching Calcutta on 25 May 1942.

The Air Defence Brigade of MNBDO 1 arrived in Colombo just in time participate in the anti-aircraft defences when the Japanese Nagumo Task Force rampaged through the Indian Ocean, their navy dive-bombers sinking the carrier *Hermes* and the cruisers *Cornwall* and *Dorsetshire*.

A successful amphibious operation was however mounted with the occupation of Vichy French-held Diego Suarez on the northern tip of Madagascar. Although under command of the RM Divisional HQ (Maj-Gen R G Sturges), all of the troops involved were army brigades, the 101 and 102 RM Brigades initially nominated were subsequently withdrawn because of a lack of administrative backing. However a 50-strong party of Royal Marines from the flagship *Ramillies* played a key role in the capture of the town. Capt Martin Price led his detachment in a daring flanking attack after being landed from the destroyer *Anthony* at Antsirane on 6 May.

The first Royal Marines Commandos were formed that year, after the initiative had, strangely enough, been left for the army for the first part of the war. 'A' Commando (formed from home based volunteers) and 'B' Royal Marine Commando (formed from 8th Battalion) soon became 40 and 41 (RM) Commandos. On 19 August 1942, 40 Commando saw its first action in the abortive raid on Dieppe. Lt-Col J Picton-Phillips showed no fear under heavy fire when he stood up and, with his white gloves, signalled the second wave of landing craft to turn about after the first wave had been decimated on the beaches. Despite heroism of this kind, the raid as a whole was a costly failure, with the marines losing half their strength. Inept and pathetic British propaganda tried to turn it into a success. Nevertheless, many useful lessons were learnt which were to save lives on subsequent operations, in particular the Normandy invasion two years later.

What was an undeniable success was the clandestine work done by Major H G 'Blondie' Hasler and his team under the misleading title of the Royal Marine Boom Patrol Detachment. They developed craft and techniques for small boat operations which were highlighted by the famous 'Cockleshell Heroes' raid in December 1942. Landing from the submarine *Tuna* off the mouth of the Gironde river in five 2-man canoes (another was split in launching), they paddled over 60 miles to German merchant ships moored alongside. Only two of the crews reached Bordeaux, but they managed to damage five ships, two of which subsequently sank. The euphoria of success was tempered by the fact that only Hasler and Marine W E Sparks returned to England. All six of those captured were shot by the Germans despite wearing Royal Marines flashes and badges of rank.

Other amphibious operations by the Corps featured in 1942 but the RM Division was never used in combat and grew restless. Churchill growled in June 1942 'what are the plans for its employment?' In fact there were none. The HQ, 101 and 102 RM Brigades had been held ready for the seizure of the Canary Islands should Gibraltar have fallen to a combined German/Spanish assault, but nothing had happened. When the RM Division lost its operational role, 9 and 10 RM Battalions replaced army units and some supporting arms were raised. They were offered for the North African landings, but were not used there either. Nor did the two MNBDOs fare any better, MNBDO 1 under Maj-Gen WBF Lukis, as we have seen, became split into two MNB Brigades. The planned assault on the Arakan for late 1943, for which Brig V D Thomas's brigade had started amphibious training, did not materialise due to lack of resources. MNBDO 2 moved to the Mediterranean, some units backing up the invasion of Sicily in July 1943. This enforced idleness of highly trained and skilled men was a marked failure of the Admiralty, especially as the war was developing into one of great amphibious operations all over the world. The very force, with nearly half of the Corps strength of 55,000,

Lt-Col H G 'Blondie' Hasler and Capt J D 'Jock' Stewart of the Royal Marines Boom Patrol Detachment practise canoe launching off Eastney beach in 1941. These were the forerunners of the present day SBS.

The only known photograph of seven of the marines who took part in Operation Frankton, *the Cockleshell raid on enemy shipping up the river Gironde to Bordeaux in December 1942; (from left to right), Cpl G J Sheard, Mnes W E Sparks, R Ewart and Cpl A F Laver (far side), Sgt S Wallace (left) and Mne J Conway (right). Of the five two-man canoes launched from the submarine* Tuna, *only two reached their target damaging four merchant ships with limpet mines. Only Lt-Col 'Blondie' Hasler and Mne Bill Sparks escaped, the remainder being either drowned or shot by the Germans despite wearing uniform.*

and with the most expertise in this role, was standing on the sidelines.

Eventually the Admiralty came to its senses, helped by some severe kicks in the right places by Admiral Lord Louis Mountbatten, then Chief of Combined Operations, who brought this sad state of affairs to the ears of the Chiefs-of-Staff Committee. Despite opposition from within the Corps, they approved the break-up of the RM Division and the two MNBDOs in April 1943 and the formation of more RM Commando units. This marked a decisive watershed in the history of the Corps, both in its meaningful contribution to the war effort and in shaping its future role down to the present day. It now became a Commando force as well as a ship-borne force. This was just as well, for as the surface fleet faded away in a deluge of post-war parsimony the Corps still had an important and decisive role which has ensured its survival. There was also a desperate need for men skilled in ship-handling to crew the ever-increasing numbers of specialist landing craft being built for the great amphibious operations being planned. These included landing craft being fitted out as shallow-draft support vessels with 4.7-inch guns, Bofors, Oerlikons, mortars and machine-guns able to bring powerful close-range fire-power right into the beach-heads. Royal Marines were the ideal candidates for such vessels, recalling as they did the bomb-ketches and mortar vessels of the Crimean War and before. When, in August 1943, an Admiralty Fleet Order was published saying that Royal Marines officers could command major landing craft, the 'Sailor' status of the Royal Marines 'Soldier and Sailor too', was finally fully confirmed. Another spin-off from this re-organisation was the formation of a Royal Marine Armoured Support Group equipped with 'Centaur' tanks, which could fire from landing craft approaching the beaches and subsequently operate ashore. Thus a confident and determined mood developed in the last two years of the war, the Corps regaining its self-respect and adding many new laurels to its already enviable reputation.

While the two existing commando units, 40 and 41

(RM) Commandos took part in the invasion of Sicily in July 1943, others were formed and in turn passed through the legendary Commando Training School at Achnacarry, Scotland. Thus 42 Commando (ex-1 Battalion), 43 (ex-2 Battalion), 44 (ex-3 Battalion), 45 (ex-5 Battalion), 46 (mainly ex-9 Battalion), 47 (mainly ex-10 Battalion) and 48 (ex-7 Battalion) all passed through basic commando training. The RM Commandos were brigaded with army commandos, with 41, 45, 46, 47 and 48 preparing for the Normandy invasion; 40 and 43 operating in the long campaign up the Italian mainland and later in the islands off the Dalmatian coast, while 42 and 44 headed east to Burma via India. Not all the existing units were capable of conversion. Indeed their skills had been honed to perfection and therefore the wise decision was made to use them as they stood and not to disband them. Thus the 5th RM AA Brigade was formed from the 1st and 3rd Heavy Regiments, 4th Light Regiment, HQ and Ops from the disbanding MNBDOs, combining them under the command of Brig J E Leech-Porter for service in Europe in support of the army once ashore.

In Sicily, 40 (RM) Commando (Lt-Col J C Manners) and 41 (Lt-Col B J D Lumsden) went ashore on 10 July 1943 against fairly feeble Italian opposition. They landed on the southern tip of the island, between Punta Ciriga and Punta Castellazzo, on a beach that was nicknamed 'Commando Cove'. Two months later, at Salerno on the Italian mainland, it was the Germans who held the ground and the fighting was somewhat grimmer. 'A' Troop was heavily pressed, but 'Q' Troop led by Capt Martin Scott retrieved the situation. Both Scott and Lt David Lloyd were killed, whilst Lt Peter Haydon, although seriously wounded, killed at least ten of the enemy with the rifle of a dead marine. For his gallantry he received the DSO: at the age of 19 he was the youngest ever Royal Marine to receive this award. When,

The Japanese Navy carried out many air raids on the main British Headquarters in Ceylon. A 3.7" heavy anti-aircraft gun of 'C' Battery, 1st HAA Regt (MNBDO 1) near Colombo soon after a Japanese air attack on 6 April 1942. The Regiment subsequently saw service in India in 1943 and North-west Europe in 1945.

Royal Marine Engineers (RME) were formed on 19 March 1940, most of the officers being civil engineers. The force rose to 8,000 all ranks by 1945. Those employed in the Far East included building an MTB base in Ceylon, the Gan airfield on Adu Atoll and repairing docks at the liberated ports of Akyab, Rangoon, Singapore and others. Here they support 3 Commando Brigade's landing at Kangaw in Burma re-inforcing the Chaung beachhead.

because of heavy losses, 41 Commando was withdrawn to England, 43 Commando (Lt-Col R W B Simmonds) took their place. They went ashore on the northern perimeter of the beachhead and took the vital Vietri Pass, withstanding a series of German counter-attacks during the next week. Meanwhile 40 Commando had landed at Vibo Valentino, Calabria, in southern Italy, and had moved up to take Termoli on the east coast on 1 October, when a flank attack by 'A' Troop led by Lt D P Hill captured the German HQ after hand-to-hand fighting. 'P' Troop (Capt Marshall) broke a German counter-attack there the following day.

By January 1944 both 40 and 41 were operating out of the Naples area, and on the 20th 43 Commando went ashore with the leading elements to take an Anzio beachhead where they fought for three long days against fierce German opposition. There was to be no break-out this time, and the Commando was withdrawn and replaced by 40 Commando. More heavy German counter-attacks and close bombing were a feature of their three-weeks stint in the bridgehead. Meanwhile 43 Commando saw action again, this time on the Garigliano River in February.

Both units then moved over to the Adriatic to co-operate with Tito's partisans, being based on the island of Vis. In a 2nd Special Service Brigade attack on the island of Brac, opposite the port of Split between 1 and 4 June 1944, the Commanding Officer of 40 Commando, Lt-Col Manners, already seriously wounded, was killed leading his commando in a counter attack against strong German positions. 40 Commando then moved south to Albania and thence to Corfu, whilst 43 Commando continued to support Tito. But due to political pressure at the highest levels, all British forces had to withdraw from Yugoslavia in January 1945.

By February both Royal Marines Commandos were back on the Italian mainland taking part in the long, dour struggle up the coast towards the Austrian border, crossing river after river against stubborn opposition. At the crossing of the Reno river, near Lake Comacchio on the east coast, on 1 April 1945, 43 Commando, by now commanded by Lt-Col I H Riches, played a significant part as they moved towards the Valetta Canal. It was here that Cpl Tom Hunter earned for the Corps its solitary Victoria Cross of World War II, when 'C' Troop found itself pinned

Marines of 47(RM) Commando embark in LCTs at Ostend for the assault on Walcheren Island on 1 November 1944. Those in the centre are sitting on one of the 'Hobart Funnies', specialist tanks of the 79th Armoured Division, commanded by Maj-Gen Sir Percy Hobart. These 'Funnies' included Armoured Ramp Carriers (ARKs), Flame Throwers, Carpet Layers, Flails and DD Swimming tanks though all of these were not used in this operation to open up the River Scheldt and the port of Antwerp in which three Royal Marines Commandos (41, 47 & 48) took part.

down by heavy machine-gun fire. Hunter, completely oblivious of his own safety, charged enemy positions firing his Bren Gun from the hip until he was cut down. His action cleared the way for the unit to advance and take the crossing successfully. The final mopping up and advance continued on the 6th, the Commando pushing on until the ceasefire sounded. In September 1945 on their return to UK, 40 Commando led for the last 9 months by Lt-Col R W Sankey, absorbed 43 Commando, but was itself disbanded in 1946.

Retracing our steps to England, the early summer of 1944 saw the tremendous build-up for Operation *Overlord*, the invasion of Normandy on 6 June. There were four RM Commandos in the initial assault. 41 Commando (Lt-Col T M Gray) and 45 Commando (Lt-Col N C Ries) landed on Sword Beach, 47 Commando (Lt-Col C F Phillips) on Gold Beach and 48 Commando (Lt-Col J M Moulton) on Juno Beach, with 46 Commando (Lt-Col C R Hardy) in the close follow-up forces. In addition Royal Marines manned nearly two-thirds of the assault landing craft, while all the larger bombarding ships had guns manned by Royal Marines. It is estimated that over 17,500 took part in the D-Day landings and immediate build-up.

The Commandos ashore on D-Day itself were 45 Commando which took the River Orne crossings reaching as far as Merville, 41 Commando who captured Lion-sur-Mer, 47 Commando who had a fierce battle for Port-en-Bessin and 48 Commando who took the strongpoint of Langrune, before advancing to overrun Douvres radio station. 41 Commando lost Major Barclay in the attack on Lion-sur-Mer and suffered many casualties, including the Commanding Officer who was wounded. 48 Commando landed under intense German fire at St

LCF 32 in the English Channel on her way to the Normandy beaches on 6 June 1944. These Landing Craft Flak (Flakkers) were based on the LCT hull, had a crew of two naval officers and ten seamen. The RM detachment who manned the 2-pdr pom-poms and 20mm Light Anti-aircraft guns (usually totalling 12) was two officers and 48 NCOs and men.

C Troop, 43(RM) Commando, in Italy in 1944 commanded by Capt (later Maj-Gen) R B Loudoun. Cpl Tom Hunter, who won the only Corps VC of World War II in the operations around Lake Comacchio, is pictured in the 4th row from the front, 4th from the right. His citation is in Appendix C.

Lt-Col T M Gray, who was wounded leading 45 Commando on D-Day (6 June 1944), receiving the DSO from Field Marshal Sir Bernard Montgomery at Kiel on 1 June 1945. Royal Marines wore the numeral of their Commando above the shoulder 'flash' and the Combined Operations badge below.

Aubin-sur-Mer, and took appalling 30 per cent casualties from the very active enemy fire. Their objective was the German redoubt at Langrune, and it took two days of heavy fighting with tank support from 2nd RM Armoured Support Regiment to succeed. They lost over half their men in achieving this victory.

On D-Day +1, 46 Commando, serving under the Canadians, landed at Bernières, capturing the strongpoint of Petit-Enfers after heavy fighting, before pushing inland to Delivrande and Douvres. 47 Commando attacked the vital fishing village of Port-en-Bessin, which the Germans had heavily fortified. The planned attack was by way of a flanking assault through Le Hamel but, met by heavy defensive fire, the fourteen landing craft took casualties and after several had been sunk the remainder diverted to the east. 47 Commando lost a third of its specialist equipment before it got ashore and the Germans still held La Rosière village, their planned assembly point. On 7 June, after a heavy naval bombardment, the attack went in but was held as the Germans counter-attacked. Only a daring surprise night attack by Capt T F Cousins turned the situation and the strongpoint fell on the following day. After the breakout, the Commandos operated north towards Dunkirk and Belgium while landing craft manned the 'Trout' line, off the beachhead, to protect the mass of build-up shipping against the possibility of one-man human torpedo attacks and the like.

As the initial Allied impetus petered out during the autumn, Antwerp and the Scheldt again resumed their age-old importance. The Allies badly needed the port to resume their momentum and the Germans knew it. Thus the attack on Walcheren Island was mounted. This resulted in one of the bloodiest fights of the campaign. The price was paid in full measure by Royal Marines in terms both of landing craft and of commandos, who bore the brunt of the fighting from 1 November onwards. The commando units involved under the command of 4 Commando Brigade (Brig B W Leicester) were 41 Commando (Lt-Col E C E Palmer), 47 Commando (Lt-Col C F Phillips) and 48 Commando (Lt-Col J M Moulton). They met heavy resistance from the German guns ashore and suffered

A Centaur tank of the RM Armoured Support Group in France in 1944. The Group was raised in the spring of 1944 for operations in North-West Europe. It consisted of two Armoured Support Regiments equipped with 20 Sherman and 80 Centaur tanks. The latter had a 3.7" gun/howitzer, and they fired in pairs from specially adapted LCT Mk 5s on the run in to the beach on D-Day, subsequently supporting the forces ashore.

heavy losses. It took three days of severe fighting before Walcheren fell. Meanwhile the 5th RM AA Brigade (Brig J E Leech-Porter) saw some unusual action when they had to engage German midget submarines with their anti-aircraft weapons at Ostend!

45 and 46 Commandos continued to feature in the final advances through the Netherlands and into Germany proper, with 45 fighting a particularly hard battle at Montforterbeek Brook on 23 January 1945. Here, one of their RAMC orderlies, Lance-Cpl Eric Harden won the VC for returning through heavy enemy fire a third time to rescue wounded men, when he was mortally wounded. Later they took part in the Rhine crossing at Wessel on 23 March. By 3 April they had reached Osnabruck from whence they returned home in June. 46 Commando followed a similar course, also taking part in the Rhine crossing, holding a bridgehead against strong opposition. Then they crossed the Aller and Elbe the following week to reach Neustadt, near Lubeck on the Baltic, as the war in Europe came to an end. They returned to UK on 8 June, a month after VE-Day. Nearly 1,000 Royal Marines had been killed in the north-west Europe campaign.

For those Marines in the Far East, the war was far from over. The remaining enemy was now Japan. In Burma, Akyab was stormed by 42 Commando (Lt-Col H D Fellowes) and 44 Commando (Lt-Col F C Horton) after the units had spent some time carrying out small but important raids on the Arakan coast. At the end of January 1945, 3 Commando Brigade Royal Marines (Brig C R Hardy) was involved in fierce fighting at Kangaw village's Hill 170 on the Myebon peninsular, 70 miles south of Akyab, and held it against suicidal Japanese attacks. 44 Commando even used elephants as transport in its own operations in the Burmese hinterland. 900 Marines manned the landing craft which did such a vital job over thousands of miles of jungle-strewn waterways. They re-occupied Rangoon on 28 August 1945 after the Japanese surrender. Similarly Force *Boots* under Capt C E Boothby took Sabang island, and other Marines were present aboard the battleship *Nelson* at the surrender of Singapore.

Over in the Pacific the battleships, cruisers and aircraft carriers of the British Pacific Fleet took the fight to the enemy in 1945, bombarding factories near Tokyo and conducting many air strikes over the Japanese mainland. Royal Marine pilots featured prominently in these air operations. Amongst them Lt Col R C Hay acted as Air Co-ordinator during the carrier strikes on oil refineries near Palambang in Sumatra on 24 January 1945 and another famous flyer Major V B G Cheesman led 1770 Squadron's Fairey *Fireflies* in strafing attacks from Sumatra,

Survivors of LCG(M) 101 (Lt George Flamank RNVR) which supported the landings at Walcheren Island on 1st November 1944. These craft mounted two 17-pdrs with RM gun crews forward and engaged targets during the run in to the beach. After knocking out her first bunker 101 was hit by a German 8.7" shell holing her astern. She rolled over and sank with only half of her crew surviving.

Marines of the 11th Battalion practice landings near Colombo in preparation for the assault on the Japanese occupied territories in the Far East. This battalion was the land defence force of MNBDO 1.

President Harry S Truman, who succeeded Franklin D Roosevelt as President of the United States of America on 12 April 1945, visits the battle-cruiser HMS Renown *in late 1945 and is received by the Royal Marines Guard of Honour on the quarterdeck.*

Indian troops embarked in Landing Craft Assault (LCAs) manned by Royal Marines, moving up the Kaladan river after landing at Akyab in 1945.

Royal Marines escort the Vicar of St Mary's Church, Chatham in a tradition that echoes back to the 17th Century. During the reign of the late Stuarts and early Georges, rectors found it necessary to take the church silver home for safe custody. About the middle of the 18th Century, the incumbent sought protection from the Royal Marines, who worshipped at this church, and they duly provided an escort. On a Sunday in November each year, this duty was still carried out until the Corps withdrew from Chatham, and the custom is now in abeyance until such time as an association between the church and the Royal Marines is restored.

British LCMs (Mk 1) land unopposed at Rangoon in May 1945. An earlier landing by LCAs had taken place 18 miles south of the city, but most of the troops were subsequently transported by LCM. These early Landing Craft (Mechanised) of which 500 were built, were designed to carry a tank or truck of less than 16 tons, two anti-tank guns or six jeeps. They each had a crew of six Royal Marines.

Vice-Admiral Lord Louis Mountbatten reads the Japanese surrender document on the steps of the Civic Buildings opposite Singapore padang. The Royal Marines are from 35 RM Provost Company.

supporting the American landings on Okinawa, and finally to the Tokyo plain. Cheesman's log indicates that these actions went on for 62 days. The part played by the 31 Royal Marines Pilots (and 2 observers) who saw action in World War II is reflected in the award to them of 5 DSOs, 1 OBE, 9 DSCs and 1 MBE. Incidentally, three Royal Marines qualified for the Battle of Britain 'bar' in the early days of the war.

The setting of the Rising Sun was marked at the surrender ceremony in August 1945 when the Union Flag was officially hoisted ashore at Yokohama, most fittingly, by Major P L Norcock, whose grandfather had carried the Colour with the Royal Marines battalion which landed in Yokohama in 1870.

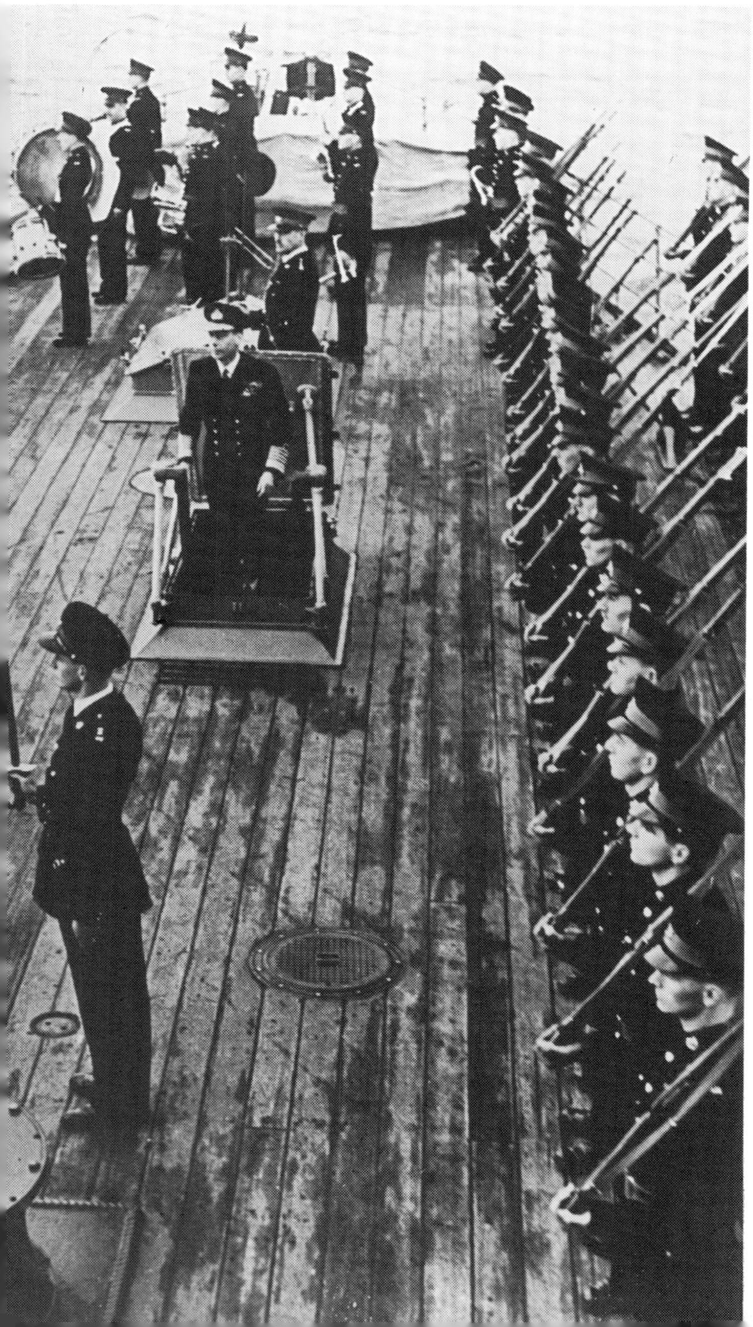

Their Majesties King George VI and Queen Elizabeth visited Eastney in January 1941. On parade were some of the Special Reservists who trained there before the Reserve Depot at Exton (later Lympstone) was opened in April 1940.

Left:
HM King George VI being greeted by a Royal Guard of Honour with band aboard the battleship HMS Malaya *in 1943. Marines at sea continued to wear blue uniform throughout the war but on land some HOs were only issued with khaki. On such Royal visits, as happened when the King visited another battleship HMS* Duke of York *during wartime gunnery exercises, he was shown 'Y' turret (the Royal Marines turret), 'as being the cleanest in the ship.'*

Opposite page
Above:
Capt J A Taplin inspects A Troop of 41 (RM) Commando prior to the Sicily landings on 10 June 1943. The SMLE with long bayonet was still being used and each man carries a toggle rope for the construction of rope bridges and climbing. From right are Capt Taplin's MOA (Marine Officer's Attendant) Mne G T Jones, Mne Parker, Cpl Parsons and CSgt H E Elsom.

Below:
Lt-Col G B Grant, commanding Force Roma *(the initial two letters of Royal Marines), interrogates Japanese staff officers on the jetty at Penang, Malaya, on 3 September 1945. Force* Roma, *formed from the marines of HMS* Nelson, Nigeria, Phoebe *and* Ceylon *detachments, landed from the LSIs* Princess Beatrix *and* Queen Emma. *They were the first British troops to occupy any part of Malaya.*

CHAPTER 10

1946-1970
WITHDRAWAL FROM THE EMPIRE

'Guard these Colours well and remember that, whatever the problem, a Royal Marine Commando is always expected to achieve the impossible.'

HRH The Duke of Edinburgh, presenting Colours to the Commandos in Malta. 29 November 1952.

This chapter could almost be headed the abandonment of the Empire. That which the Corps had done so much to build in the preceding two centuries was demolished in a few decades, but the cost of the wholesale retreat proved nearly as high as its original foundation. It was to be expected that the Royal Marines strength would fall dramatically, having reached an all time peak of over 78,000 in 1945. They had suffered nearly ten per cent casualties, indeed the Royal Marines Band Service alone had suffered 225 dead, which was the highest percentage of any service. The coming of the new, uneasy peace brought wholesale retrenchment.

The peace that followed World War II was very different from that which followed the Great War, for it was not long before the overtones of the 'Cold War' began to manifest themselves. After so long playing second fiddle to the United States in the NATO Alliance, it is difficult to realise that in 1945 Britain was still a premier world power with her Empire intact, and still the second largest naval power with over one million men under arms. However, the signs of change were already there before the war finished in the Far East. A Socialist government, elected on 26 July 1945, pledged to reduce arms spending and concentrate on post-war re-building. They were also fiercely doctrinally opposed to the whole concept of the Empire and resolved to grant independence to nations like India and Burma as soon as possible, and certainly before either country had any time to recover from the effects of five years of war. Britain's economy was in a shambles, mortgaged to fight the war and highly dependent on Lend-Lease, which the Americans, no longer our Allies but just another alien power that had already usurped our principal role in the world, were quick to cut off. Britain's reduced circumstances coincided with the difficult period of transition from Empire to off-shore island. It proved a long and arduous road.

On 8 June 1946 the Massed Bands of the Royal Marines led the Royal Navy contingent up the Mall towards Buckingham Palace during the Victory parade. Lt-Col F B Pym led 12 officers and 192 NCOs and men with the Colours of Chatham, Portsmouth and Plymouth Divisions, and were followed by contingents from Queen Alexandra's Royal Naval Nursing Service and the Women's Royal Naval Service.

There was also the dramatic run-down of the fleet, far more severe even than in the post-Washington Conference era. The traditional role of the Royal Marines afloat soon vanished as all save five of the remaining battleships were sent to the scrapyard, and even these five followed by the mid-50s. Likewise the cruiser fleet was rapidly reduced, from over seventy ships to a mere twenty. The atomic bomb cast a long shadow over conventional surface fleets. By the early 1960s the Royal Navy had sunk to fourth or fifth place in the ranks of world sea power, and still continues to shrink to this day under continued cut-backs in surface capability. The custom of carrying Royal Marines detachments, which reached a peak with nearly 400 Marines aboard a battleship in wartime, only continued on certain small frigates

The Band of the Plymouth Division, as the RM Far East Band, outside the Municipal Buildings in Singapore on 6 January 1946, as the C-in-C South East Asia Command, Admiral Lord Mountbatten, leaves after presenting medals to guerilla leaders who had fought against the Japanese in the Malayan jungles. Some of these men subsequently returned to the jungle to fight against the British during the long terrorist campaign of the 1950s in which 3 Commando Brigade took a prominent part.

and the remaining few aircraft carriers but their numbers were tiny.

That the Corps continued to have a role at all was due to the parliamentary decision of October 1945 which disbanded all army Commando units and transferred their role exclusively to the Royal Marines. Only a single Brigade was retained, 3 Commndo Brigade then in Hong Kong. The three commandos assigned to the brigade were 44 (RM) Commando (which became the new 40 Commando Royal Marines), 42 and 45. This retained the name of one commando each from the three main theatres of the war, Italy, the Far East and North-West Europe respectively. To establish a firm base for reserves in case of emergency, the Royal Marine Forces Volunteer Reserve was established three years later. The name was changed to Royal Marines Reserve on 1 October 1966, initially 1,500 strong with training centres organised in three wings, Amphibious Operations, Commandos and Sea Service. Centres were eventually established in London, Bristol, Glasgow, Liverpool and Newcastle. In 1946 the Royal Marines Association was formed, which has been stoutly maintained up to this day and shows no signs of wilting.

Under such straitened circumstances many of the old cherished traditions had to go. First to disappear were the 'Grand Divisions', although these were replaced by geographical Groups covering the same areas of Chatham, Portsmouth and Plymouth. The post of Adjutant General Royal Marines, became briefly G-O-C RM before becoming Commandant General Royal Marines in July 1945. One improvement was that he had direct access to the Board of Admiralty, while that august body was still in being although its days were numbered. With the arrival of General Sir Dallas Brooks at the top, all future drafting requirements, records, pay and NCO promotions were centralised, although Divisonal prefixes were still retained for registered numbers until 1948.

The new arrangements did not long survive the continued cut-backs, and by 1949 the total strength of the Corps was down to 10,000, of whom about 2,000 were at sea, which necessitated the abolition of Chatham Division. The passing of this proud and ancient association, dating back to 1664, was commemorated by the laying up of the old Divisional Colours in Rochester Cathedral. For a time a faint link was retained with the centralised Drafting, Pay

One of the important duties carried out on board ship was that of Keyboard Sentry, who was normally situated in the officers flat aft. Four marines would work watches at this duty, two each on alternate days. Also stowed in the after flats were rifles and pistols for use by the ship's landing party.

A bugler sounding 'Stand Easy' over a ship's tannoy system of a cruiser in 1947. Blue battledress was worn at sea as an alternative to No 3s (blues) until the early 1950s and for sea service and amphibious training ashore. Normally two buglers, one often a junior, were carried in each detachment and apart from bugling, they helped to look after the Sergeants Mess. The Bugler Branch was absorbed into the Band Service in 1978, but they continue to wear their distinctive dress cords and a thin stripe down their trousers today.

and Records Office still being based in Melville Barracks, but in 1960 this moved to Portsmouth and the last permanent link was broken. The RM Band of Western Fleet, later the C-in-C Fleet's Band, was based there for several years and there is still an active Marines Cadet unit there.

3rd Commando Brigade was actively involved in the resettlement of Hong Kong during 1946, when the transition from an almost totally Hostilities Only formation to a mostly regular force, saw large changes in personnel with many marines who had spent the war at sea joining commandos for the first time along with the first of the National Servicemen. The Brigade moved to the Mediterranean area in 1947, based on Malta as the Middle East mobile reserve.

Although pledged to withdraw from its world-wide commitments, Britain faced even more problems than ever in the post-war years. The first of these set the tone for the many that followed: this was the Palestinian emergency. It had been a long-running sore even in the 1930s when Royal Marines had to be landed to stand between Jew and Arab, but after the war, the trickle of illegal Jewish refugees became a flood. Sympathy was universally with the Jewish survivors of the holocaust, but the Arabs in Palestine, still under British mandate, naturally objected to being crowded out of their own land. As usual, the British soldier stood in the middle, while the politicians argued at the newly-constituted United Nations. As the sailing of clandestine ships packed with refugees escalated, helped by Americans out for a quick profit, others in Palestine were loud in their condemnation of Britain's attempts to draw a reasonable line. Soon new factors arose such as terrorist bombing and murder squads. These quickly brought indiscriminate death and injury to British policemen, service-men and innocent civilians alike, and set a pattern now so well established. The Royal Marines played their full part in this unhappy episode, not resolved until 1948 when the independent state of Israel was founded. There was 40 Commando serving ashore at Haifa, guarding the docks and installations from sabotage, whilst trying to keep the opposing factions apart. They patrolled the terror-strewn streets and also acted as boarding parties from the destroyers of the Mediterranean Fleet, carrying out the thankless task of trying to turn back the armada of refugee ships with minimum violence.

Meanwhile 42 and 45 Commandos initially saw service in Libya, being based at Tripoli during the post-war Four Nations Committee which decided the fate of the former Italian North African colonies. When the British withdrawal from Palestine was announced, they joined 40 Commando there in April

The Royal Marines orderlies who accompanied the Royal Tour of South Africa in HMS Vanguard, *the last British battleship, in 1947. With HM King George VI and HM Queen Elizabeth, the Princesses Elizabeth and Margaret Rose are Sgt A Rendle, C-Sgt J Burt, Sgt D Costley and Sgt R Fry.*

Sir Malcolm Sargent, who was appointed Honorary Adviser in Music to the Royal Marines in 1949, visits the RN School of Music at Burford, Oxfordshire on 22 January 1947, where it was situated from May 1946 until January 1950 after earlier wartime evacuation to Scarborough with the Boy's Wing in the Isle of Man. Sir Malcolm is accompanied by Capt Herbert Kenward, the Director of the School who had joined as a Band Boy in 1908.

1948, providing yet again, the final covering force. The latter transferred to Cyprus where camps of illegal immigrants had been established and were causing more difficulties, while 42 and 45 served alternately in Malta and the Suez Canal Zone. Here they protected British interests and installations particularly on the telephone and power cable tracks, which were prone to sabotage. In April 1949 45 Commando were sent hurriedly to Aqaba in Transjordan where Jewish forces threatened the port. But no trouble materialised and they returned to their base in Suez two months later.

However the rise of violent Nationalistic forces in Egypt and the success of Israeli terrorist groups who were resolved to follow their example, saw a spread of urban terrorism to Cyprus, which was reflected elsewhere in the world. In China, the civil war between the Nationalists and the Communists reached a peak in 1949, terminating in the expulsion of the former under Generalissimo Chiang-Kai-Shek to Formosa. Hong Kong was severely threatened and the British government pledged to fight to defend its integrity. At very short notice the whole of 3 Commando Brigade were rushed there, Brigade HQ, 40 and 42 Commandos sailing from Malta on 21 July in HMT *Georgic*, picking up 45 Commando at Port Suez on the way. They arrived in the heat of Hong Kong on 13 August to be greeted by the pipe band of 1st Battalion Argyll and Sutherland Highlanders, one of their 'chummy' army regiments, their musical salute being returned by 3 Commando Brigade's own Royal Marines band. This band had been formed 25-strong early in 1948. In the event the Chinese drew back from invading the Crown Colony, but the Brigade's task was the internal security of Victoria Island and the hundreds of outlying islands that made up Hong Kong. But there were other commitments in the Far East to occupy the Brigade over the next few years. As we shall see later, communist terrorism occupied 3 Commando Brigade in Malaya for the next 2 1/2

The wartime Commando School was based at Achnacarry near Spean Bridge in the highlands of Scotland and moved for a short time to Towyn in North Wales. In 1947 it was established at Bickleigh, near Plymouth on the edge of Dartmoor which has made an ideal training area ever since. Thousands of recruits and most National Servicemen carried out their initial commando training in Bickleigh Vale where the River Pym provided a cold welcome for many.

Left:
On 8 June 1948 the Royal Marines provided a Guard of Honour outside the Guildhall in London when Lieut Philip Mountbatten received the Freedom of the City of London. The guard was provided from Chatham Group and carried the Regimental Colour of Chatham Division.

A dedication service conducted by the Chaplain of the Fleet at the RN School of Music at Burford on 1 June 1948, to mark the presentation of fourteen Memorial Silver Trumpets. These were bought by private subscription in memory of the 225 musicians who lost their lives in World War II. This was the largest percentage fatal casualties in any branch of the Services. These trumpets are now on display in the Royal Marines Museum.

King George VI dined with 300 officers of the Corps at the Savoy Hotel, London, on 21 December 1949. During the dinner the tradition of lighting the King's Candlesticks in Officers Messes was established. The Globe & Laurel *reported 'This simple piece of ceremonial seems particularly appropriate. A new flame of loyal tradition has been kindled which will fire young spirits many generations ahead.'*

(See Appendix A)

years and in April and May 1950, the Commandos were sent to one of the worst trouble spots, the state of Perak, in north Malaya.

Therefore no Royal Marines were initially on hand when the communist North Korean troops crossed the 38th parallel to invade South Korea on 25 June 1950. Due to the fortuitous absence of the Russian delegate from the Security Council chamber of the United Nations, for the first and only time in its life so far, it was able to take concerted action against communist aggression, and a resolution was passed for the despatch of UN troops to aid the Seoul government. The first marines to see action were those in the warships of the Far East Station, light fleet carriers and 6-inch gun cruisers, who were made responsible for the west coast of Korea and who conducted several bombardments in support of the small Pusan perimeter still held by friendly forces. The British C-in-C, Admiral Sir Patrick Brind, offered a small raiding force against communist lines of communication and this was accepted. However by this time 3 Commando Brigade was already heavily involved in Malaya and a new force was required.

A special Royal Marines Commando was swiftly raised at the Commando School, Bickleigh, on 16 August 1950 to be commanded by Lt-Col D B Drysdale. It was designated 41 (Independent) Commando Royal Marines and consisted of volunteers from UK establishments who were flown out to Japan via Rangoon by BOAC in plain clothes. Here they were joined by volunteers from the fleet who were already in training and by a re-inforcement draft from Malaya, which was diverted via the Philippines from the Troopship *Devonshire*. The unit was clothed and equipped by the Americans at Camp McGill, Takehama, and at the most it consisted of 300 all ranks.

They soon began to conduct a number of vital coastal raids in conjunctions with the US Army Special Raiding Forces. *'Poundforce'* of 14 men under Lt E G D Pounds carried out a diversionary raid on the west coast on the night of 12/13 September and on 17th joined the dramatic Inchon landings which temporarily reversed the communist tide, penetrating as far as Kimpo airfield.

The Commanding Officer accompanied 'A' & 'B'

Royal Marines land from HMS Devonshire *at Grenada, West Indies on 2 February 1951 when there was rioting in the British colony. They relieved civilian police in guarding vulnerable points with sections at the Princess Alice Hospital and Pearl's Airport and remained ashore for 12 days. This landing of a detachment from an HM Ship was typical of a number carried out in support of the civil power during the post war years.*

HM Queen Elizabeth, the Queen Mother unveils the Commando Memorial at Achnacarry on 27 September 1952. This imposing statue, with three individual commando figures 9' 4" high, stands over 17' overall, and is the work of the sculptor Scott Sutherland. It commands a magnificent view of the River Spean valley, Ben Nevis and the western end of the Great Glen, familiar training grounds of the wartime commandos. At the ceremony can be seen Lord Lovat, Maj-Gen R E Laycock (a wartime Chief of Combined Operations) and Col C E Vaughan (first Commandant of the Commando Basic Training Centre), while the 5,000 spectators formed the largest gathering in the district since Prince Charles Edward Stuart raised the standard in 1745.

Troops during a raid on 2 October on the east coast railway when they planted limpet mines after landing from the US submarine *Perch*, losing their first man killed in the process. 125 men from 'C' & 'D' Troops under command of Major D L St M Aldridge landed from US destroyers *Bass* and *Wantuck* on 5/6 October successfully laying charges in tunnels, on railways bridges and culverts, again for the loss of a corporal. 41 Commando then returned to Japan for further training on the slopes of Mount Fujiyama before being attached to the 1st US Marine Division, the two Marines Corps fighting together as they had done half a century before in Peking.

The Coronation of HM Queen Elizabeth II on 2 June 1953. The Royal Marines processional contingent of 13 officers and 132 rank and file under the command of Major P G Solbe, headed by the combined bands of Plymouth Group and the Royal Marines School of Music led by Drum Majors C Bowden and L Keefe, round the Victoria Memorial opposite Buckingham Palace. In addition the Corps provided a Royal Guard outside Westminster Abbey under the command of Major A J S Crockett, with the Band of Portsmouth Group, led by Drum Major R G Knox in attendance.

The US force landed on the east coast of Korea at Hungnam and had advanced up the Songchon river via Hamnung, Sudong and Hagaru to the Chosin reservoir as the North Korean forces fell back all along the line, almost to the Chinese border. Hopes were high that the war would soon be over and 'Home by Christmas' was again often heard. This euphoria quickly evaporated when the Chinese Communist armies were ordered into combat in October and it was soon apparent that their overwhelming numbers had been concentrating on the Yalu river in huge strength. Meanwhile 41 Commando had put its head into the lion's mouth, arriving in Hungnam on 15 November, where they were issued with American winter clothing. But they still retained their green berets and most of them wore their British boots as well! The role of this 235-man force was to act as an additional reconnaissance Company on the US Marines flank during the planned advance from Yudam Ni to the border. But the plan was abruptly overtaken by events.

The advance began for 41 Commando on 29 November and after heavy fighting, they had only penetrated four miles beyond Kotori on the first Next day they were ambushed in a narrow defile on the way to Hagaru and the column was split in two. Less than 100 of the Marines, many suffering from frost-bite and others, including Drysdale, wounded managed to fight their way through to Hagaru where they formed the garrison reserve. 61 marines were battle casualties or had been forced to surrender after

41 Independent Commando was formed in UK on 16 August 1950 for service with the United Nations Forces in Korea. After carrying out raids on the North Korean coast, they took part in the epic withdrawal from the Chosin Reservoir as part of the US 1st Marine Division. They subsequently raided the east coast railway at Sorye Dong in April 1951 and here are seen in defensive covering positions. Equipped with US weapons - here are 2.65mm rocket launchers (the original Bazooka) and the 75mm Recoiless rifle - they were kitted in American uniforms but retained their green berets and mostly their boots and puttees.

Capt Pat Shuldham briefs members of B Troop, 41 Independent Commando aboard the USS Marion *prior to a daylight raid on enemy supply routes 8 miles south of Songjin on the east coast of Korea on 7 April 1951. The mixture of British and American clothing worn can be clearly seen.*

fighting in small isolated groups during the night. One group of 25 under the assault engineer officer, Capt P J Ovens, managed to return on their own through hazardous and mountainous country in the bitter cold to Kotori.

In the Hagaru-ri trap, the remnants fought on from 30 November until the breakout. One incident worthy of their fighting spirit was when Lt G F D Roberts led a patrol from 'B' Troop in a vital night counter-attack across completely unknown country, to plug a gap in the defences. About 25 Marine were flown out by 5 December, but orders for the US Marines to abandon their heavy equipment were scorned by the Corps commander, Maj-Gen Smith and the US Marines announced their intention of 'Advancing to the South'. A later motion picture of this epic withdrawal summed it all up in true Hollywood style, 'Retreat Hell - we're advancing in another direction!' The withdrawal commenced on 6 December and with them went the Royal Marines after Drysdale had carried out a formal morning inspection. The Royal Marines custom of shaving daily despite the freezing weather was initially greeted with derision by the Americans, but they eventually conceded that there was something in these outward signs of self-discipline. The depleted unit acted as rearguard and they had to fight their way through an estimated seven Chinese Divisions before they reached the coast again.

On 8 December Kotori was reached and Ovens's 25 men re-joined the unit to bring the strength up to 150. The most hazardous part of the battle now lay through the Funchilin Pass. 41 Commando moved out in a blinding snowstorm during the afternoon to hold some vital high ground. It was still 23 miles to the Hungnam bridgehead and by 11th they finally reached comparative safety, but 93 marines had paid the price. By Christmas 41 Commando was not home but at Masan, down the coast. The Commando was then withdrawn to Japan to re-train, re-equip and await reinforcements.

Col Drysdale was to comment on the fighting spirit of the US Marines thus:

> Let it be said that the admiration of all ranks of 41 Commando for their brothers in arms was and is unbounded. They fought like tigers and their morale and *esprit de corps* was second to none.

A cartoon of E G Oakley-Beuttler which was used as a Christmas card by the Commando School, Bickleigh in 1949.

On their part the Americans awarded 41 Commando a Presidential Unit Citation which was subsequently received by the Captain General from the US Ambassador in 1957.

Whilst awaiting specialist personnel from home, a further raid was carried out at Sorye Dong on the morning of 7 April 1951. 277 officers and men were landed from American landing ships and destroyed the embankment carrying the coastal railway with charges placed in boreholes by the assault engineers, aided by the MT section. Although protracted peace talks were under way (and still are!) 41 prepared for larger operations to establish a forward base on Yo Do Island in Wonsan harbour, some 60 miles behind enemy lines. Accordingly, preliminary landings were made on small islands in the area, 'B' Troop under Capt E T G Shuldham, taking Modo on 9 August while 'D' Troop (Capt A Stoddart) re-occupied Taedo Island. To the north Hodo Pando island was raided by canoe parties from 'B' and 'C' Troops between 11 and 29 August.

Lt-Col F N Grant took over command on 15 October and on 30 November, a landing was made between Songjin and Hungnam which was opposed with the loss of three wounded. Another raid the next night suffered a similar fate. Further raids were planned but in late December, and completely out of the blue, 41 Commando was ordered to pull out altogether. Their final raid, Operation *Swan Song* was led by Lieut J R H Walter and Sgt R R Dodds of 'D' Troop who destroyed some enemy craft at Changguok on the north west coast of Hondo Pando. 41 Commando - embarked on 22/23 December and sailed for Sasebo in Japan, from where they sailed for home in the *'Empire Orwell'*, finally disbanding at Bickleigh on 22 February 1952. They left behind in Korea 31 dead and 17 POWs, who were not released until 1954. Seven others died in captivity. Despite intensive 'brain-washing' used here for the first time by the communists, only one man, Marine A M Condron, was so affected that he stayed behind. The remainder resisted these vile and insidious interrogations.

Many headaches had followed the easy Japanese conquest and occupation of former Colonial territories in that area. The seeds of discontent with white rule, never far below the surface, had been watered by the British defeat by an Asian nation and

Cricket has played an important part of Corps sport and this scene of a match on South Green, Deal, will evoke memories of the road which runs at right angles to the wicket only 30 yards behind the bowler and 20 yards inside the boundary. The Corps cricket trophy, the Mead Cup is now competed for annually between Royal Marines units, although it was originally played for between the RMLI Divisions. It was presented in 1920 by Mr A G Mead in memory of his son Capt Claud F Mead RMLI who was killed in action in Gallipoli on 13 December 1915.

At the State Opening of Parliament on 4 November 1952, former Royal Marines of the Honourable Corps of Gentlemen at Arms and the Queen's bodyguard of the Yeomen of the Guard are pictured together. From left: Lt-Col K E Previtte OBE, Lt N A Finch VC, CSM R E Payne DCM, RVM, MSM and Maj-Gen A R Chater CB, DSO, OBE.

blossomed post-war. The Communists cells, active in every country, were quick to latch on to any such manifestations and turn them to their own advantage. Thus the Malayan terrorist groups, Lie Ton Ten and Min Yuen grew bolder with foreign financial aid and backing until, by 1950, a full-scale war was being conducted in the Malayan hinterland. They combined under the Red Banner as the 'Malayan Races Liberation Army', formed from the remnants of the wartime 'Malayan People's Anti-Japanese Army', and the forerunner of many such organisations in that part of the world.

Isolated British planters were murdered and their workers intimidated or recruited, while bloodshed and arson became rife. Eventually the ability of the normal civil defence forces to cope was questioned

and, as in Palestine, British troops had to be called in to conduct full scale search and destroy operations against dissident groups, this time deep in the jungle. The Royal Marines in Hong Kong were the closest trouble-shooters and in May and June 1950, Brigade HQ and the three commandos sailed for Malaya. 40 Commando was based at Kuala Kangsar in northern Perak, 45 Commando at Tapah to the south of the State and also responsible for the Cameron Highlands in Pahang some 35 miles up a winding, narrow and dangerous road, while 42 Commando shared Ipoh with Brigade Headquarters. The enormous area which stretched for over 8,000 square miles was one of great complexity with the jungle-clad hills crossed by small streams, combining with vast areas of lallang-covered swamps, being almost impenetrable. Patrols were sent out into the heart of these tracts, following rivers or hill-top tracks, hacking their trails inland to visit isolated kampongs and lay ambushes.

One of their main tasks in the early days was the resettlement of squatters as part of the Briggs plan, concentrating them in camps, and forcing the terrorists out of the jungle in search of food and supplies. Later deeper and longer penetration patrols sought out the bandit gangs, often being re-supplied by helicopters, who also carried out casualty evacuation. But the elusive enemy were comparatively few and only about one patrol in every twenty ever made a contact, which meant that the boredom of continuous patrolling with only an outside chance of success had to be overcome and jungle awareness had to be of the highest level. The brigade success rate of kills and captures was impressive, often relying for their information and intelligence on the British-officered Malay Police and informers. JCLOs (Junior Civil Liaison Officers), many of them Chinese and Iban (Sea Dyak) trackers from Sarawak accompanied most of the patrols, acting as guides and interpreters. For two and a half years Royal Marines operated

Major Jack Hedley is chaired from Bisley after winning the Corps Champion at Arms. During World War II Major Hedley had been awarded the DSO when he landed with 11th Battalion at Tobruk on 14 September 1942 in an ill-fated operation. Only 70 marines actually landed and Major Hedley led them ashore against fierce German and Italian opposition. Having silenced a machine-gun nest with a grenade, he killed five Italians with five shots from his pistol.

patiently in this unfriendly jungle environment, and became highly skilled at this quiet, individual and clandestine type of fighting. There was none of the wholesale firepower and blanket-bombing with napalm which characterised the Vietnam conflict a decade later. The British way was slow, controlled, highly professional and certain. Whether General Sir Gerald Templer, Director of Operations, actually coined the phrase 'Hearts and Minds' is open to question, but his wise and prudent policy of the 1950s ultimately triumphed, though 3 Commando Brigade did not remain to see it through.

By April 1952, the back of the communist bandits in the area had been broken. The Royal Marines had lost 19 officers and men killed, but had accounted for 171 terrorists killed and 49 captured in the same period winning some eighteen gallantry decorations for the Corps in the process. By July of that year, some of the brigade having moved south during their last six months, the formation returned to the Mediterranean and were once more based on Malta. It was here, on 29 November that HRH The Prince Philip, Duke of Edinburgh, presented each of the Commandos with their first Colours. Prince Philip's association with the Corps are long and deep, stemming from the time he served as a young Lieutenant aboard HMS *Valiant* of the Mediterranean Fleet. He has kept up his special relationship with the Corps ever since. Indeed he had already presented new Colours to Royal Marine Barracks Plymouth in 1951. This was fortified in his appointment as Captain General when Queen Elizabeth ascended to the throne.

By now scheming, plotting and terrorist bombing had become well-established in Egypt as the manoeuvre to throw Britain out from the strategically vital Suez Canal Zone gathered momentum. As in Palestine, Britain found few allies for her position, the Americans, Russians and lesser powers combining almost exclusively against us as so

On 14 July 1959, the Royal Marines were granted the Freedom of Portsmouth. HRH The Prince Philip, Captain General, accepted the casket containing the Freedom scroll from the Lord Mayor of Portsmouth, Lt-Col (now Sir Alfred) A L Blake, at Eastney Barracks on behalf of the Corps. Col Blake commanded 45 Commando in north-west Europe for a short time and later became the second General Secretary of the Royal Marines Association and subsequently the Director of the Duke of Edinburgh Award Scheme.

When HM The Queen visited Malta in May 1954, she received a Royal Salute from a Combined Service Parade on the Floriana parade ground. The Colours on the right are those of 45 Commando which had been presented by HRH The Duke of Edinburgh two years earlier and this was the first time all three Commandos' Colours had been on parade together.

'Who ever heard of a Marine who could march anyway!' This Giles cartoon of 1959 was presented to the RMR (City of London) when they started a marching craze which swept through the country. It resulted in Dr Barbara Moore, a walking fanatic, throwing down a challenge to the Royal Marines.

A typical guard and band parading on the quarterdeck for entering harbour during the 1950s. Here HMS Bermuda *enters a port in the south of France during an official good-will visit. A Royal Marines detachment in a cruiser was upward of 50 strong with a band of 16-20.*

Lt-Col (later Sir) F Vivian Dunn, Principal Director of Music Royal Marine presents the manuscript of a new Wren's March to the Director of WRNS, Dame Elizabeth Hoyar-Millar at the RM School of Music at Deal, watched by 2nd Officer Elizabeth Craig-McFeeley (the Unit Officer) and Capt Tony Parker (the Staff Officer) in 1960. The march is based on the traditional air 'There is a Lady Passing By'.

'It say's – "Mind you don't catch cold … signed Royal Marines"'. A Giles cartoon of 1960, the original of which is on display in the RM Museum. There has always been a friendly rivalry between the Corps and the Household Division, particularly over ceremonial matters.

often before. What brought things to a head in 1953, Coronation Year in Britain, was the overthrow of the Egyptian monarchy and the establishment of army rule. Once again it was the Commando Brigade who were sent to the Canal Zone, along with the army, to hold the fort while preparations for a final British withdrawal were made. It was an uneasy situation where a shot in the back, an indiscriminate bomb at night or another equally unexpected hazard had to be endured with little opportunity to answer back. 40 Commando who had been exercising in the Zone in February 1953, returned with 42 and 45 Commandos three months later, and their temporary deployment lasted until September/October 1954 when the British pulled out. The move of the Middle East base to the 'safer' island of Cyprus started, a move recommended during the 1930s when the MNBDOs were planned, but regrettably never implemented. Inevitably, having witnessed the success of terrorism in other Middle Eastern states, a threat immediately arose with the rise of the EOKA organisation, based on the Israeli and Egyptian terrorist gangs. 'Union with Greece' was the cry from those who had learned nothing from history and the burning of Smyrna.

The Corps retained its amphibious arm, not only in the Amphibious School at Eastney, but also by its continued presence with the RN Rhine Squadron (HMS *Royal Prince*) at Emden in Germany. Incidentally they were the only Royal Marines formation ever to serve permanently in Germany after the war. Their task from 1946 was moving allied tanks along the Rhine in case this should be needed as a holding line. They operated many types of landing and river craft and would have been responsible for demolition work in the event of an emergency. In 1953 it became a wholly Royal Marines commitment, although the Squadron had a naval captain in command, and remained so for ten years. Meanwhile in the Mediterranean an Amphibious Warfare Squadron had been created in 1952, with the headquarters ship *Meon*, the LSTs *Reggio* and *Striker* and the LCTs *Bastion* and *Redoubt* and *ML2583* which was commanded by a Royal Marines officer, the only marine entitled to fly the White Ensign. All the landing craft crews were Royal Marines and they supported 3 Commando Brigade in the Mediterranean on countless exercises.

Meanwhile further economies had been imposed on the Corps, which necessitated the return to England in 1954 of 42 Commando. An operational nucleus of the unit was stationed at Bickleigh along with the Commando School, maintaining two operational Troops and a Commando HQ, who spent the next few years exercising in Scotland, north Norway (a forerunner of the Corps' present role), Denmark and Germany. In 1957 an enlarged Troop of 100 Marines

spent over eight months based in HMS *Sea Eagle* in Londonderry on anti-IRA patrols along the Irish border. An interesting contribution to history and the reversal of the trend of dismantling of the Empire was recorded when Sgt B W Peel and Cpl A A Fraser were landed from the survey ship HMS *Vidal* on 18 September 1955 on the tiny speck of Rockall islet, three hundred miles west of the Outer Hebrides, and annexed it for Britain. This dot in the ocean was on the test path of guided missiles but the affixing of the plaque and the raising of the Union Flag by 42 Commando ensured that no 'red' cormorants or guillemots spied on the odd British rocket!

As we have seen, Royal Marines had been present in Cyprus in 1878 when the island had been declared

Cpl A A Fraser and Sgt B W Peel, with Lt Cdr D B Scott of HMS Vidal *(above) cement a plaque on the remote islet of Rockall in the Atlantic when it was annexed by Great Britain on 18 September 1955. The plaque includes the words 'The Union Flag was hoisted and possession of the island was taken in the name of Her Majesty (Queen Elizabeth II)'. The importance of the islet was apparent when under-sea oil exploration started and Britain could claim territorial rights over a large area of the North Atlantic.*

The Duke of Edinburgh's Trophy was instituted in 1956 for competition between units of the British, Commonwealth and Territorial Armies, plus the Royal Marines, in which Prince Philip held an appointment such as Colonel-in-Chief, Captain General, etc. The competition consisted of shooting and physical tests. Members of 40 Commando's winning team, led by Capt Peter Stickley (centre) survey the 1959 results in St Andrew's Barracks, Malta. In addition to this win, 41 Commando won it in 1960 and 42 Commando in 1968, while Commandos came second in 1957, 1959 and 1961. The competition ceased in 1970.

40 and 45 Commandos operated regularly during the 1950s in operations agains the EOKA terrorists in Cyprus. Here a 45 Commando patrol rests after finding an EOKA hideout in the Troodos mountains.

a British protectorate, and they had since served ashore during the frequent Greek/Turkish flare-ups. The festering of hatred between the two communities was now exploited in classic style by the EOKA, and a new reign of terrorism began on this hitherto relatively tranquil and beautiful island. Again it was 3 Commando Brigade who were thrown into the breach with their, by now, unique expertise in counter-terrorist operations. Brigade HQ, with 40 and 45 Commandos moved from Malta in September 1955 and were based on Limassol, Paphos and the Troodos Mountains respectively. For the next twelve months the young Marines, many of them National Servicemen, undertook patient but often fruitless patrols in the mountains, trying to track down the elusive terrorists who adopted normal peasant guise in the villages. When they wanted,they took up arms and killed by bullet and bomb along the infamous 'Murder Mile'. There was little tangible reward for the Marines from such operations, other than the knowledge of a job competently done, while they kept the lid down firmly and impartially on an explosive and potentially tragic situation. This was not appreciated by the Greek Cypriots at the time, but they realised it too late when the Turkish army allowed them none of the British 'kid-glove' treatment when they finally lost their patience and partitioned the island.

In November 1956 came an incident that has since become a watershed, both in British politics and in the 'modern' history of her armed forces, the emotive Suez Operation. Not since the Boer War, half a century before, had such a seemingly straight-forward policing action brought down such wrath upon Britain's collective head from other nations, and probably not since the Civil War had the nation been so decisively split on such an issue. It marked the end of any pretensions Britain may have had that she was even a second-rate military power, and led ultimately to the fall of the British Prime Minister, Sir Anthony Eden. Recent revelations from hitherto secret documents have revealed how the situation was manipulated by Britain for her own ends, and how

Royal Marines of 40 Commando storm ashore from their landing craft at Port Said on 6 November 1956. The buildings in the foreground were their first objectives while the oil refinery burning in the background was the result of air attack by the Fleet Air Arm.

Reinforcements moving in by LCAs to Port Said docks twelve hours after the main assault which was carried out across the beaches (top) by 40 and 42 Commandos on 6 November 1956. 40 Commando's first objectives, close to the Suez Canal, can be seen in the centre, and on the right is the area where 45 Commando carried out the first ever unit helicopter operation.

the government deceived the bulk of its own people. But at the time the situation seemed very clear to those ordered to carry out military operations.

The Royal Marines played the leading role in the landing. The origins of the action can be traced back to the rise of Gamil Nasser as the new Egyptian leader who following the example of Arabi Pasha in 1882, set out to convince his people of his worth by humiliating the British. In 1882 Arabi was quickly humbled but by 1956 many things had changed, not all for the better. The withdrawal of British troops from the Canal Zone and the nationalising of the Suez Canal left relations very strained indeed. Despite the fact that America enjoyed similar rights in the Panama Canal, the USA joined in the mounting pressure for Britain to withdraw. The British and French froze their Egyptian assets and re-called their canal pilots, but the only result was intervention by the Russians. Meanwhile, Israel launched armed intervention across her borders against Egypt in the Negev at the end of October. On the pretext of stepping in between the aggressors and to protect the canal, a joint occupation force was landed by Britain and France. Papers recently released indicate that there was undoubtedly collusion between Britain, France and Israel. But here we are concerned only with the military aspects.

The Royal Marines led the seaborne assault on Port Said on the morning of 6 November 1956, after a light naval and air bombardment limited by the government's belated decision to use nothing larger than 4.5-inch guns. 3 Commando Brigade, commanded by Brig R W Madoc, had been withdrawn from internal security duties in Cyprus and, reinforced by 42 Commando, brought up to full strength. Thorough training for an amphibious landing had been carried out extensively in Malta during September and October, Brigade HQ, 40 and 42 Commandos embarked in ships of the Amphibious Warfare Squadron on 30 October, sailing the following morning in one of the *'slowest Mediterranean convoys since the days of Cleopatra'*. 45 Commando embarked in the faster assault aircraft-carriers and sailed three days later.

After a drop on Gamil airfield, six miles west of Port Said, by '3' Para on 5 November, combined with a French parachute drop to the east of the Canal, the three defending battalions of the Egyptian army had withdrawn into the town with their Soviet tanks and mortars, while police and armed bands of civilians were let loose to harass the Allies in street fighting. At first light, 40 Commando (Lt-Col D G Tweed) and 42 Commando (Lt-Col P L Norcock) landed in Buffalo

Marines of 42 Commando prepare to land from an 'O' Class submarine in Gemini assault boats off the coast of Malaya in 1966.

Since World War II, the Royal Marines have provided Special Boats Sections and Squadrons. Their tasks of beach reconnaissance and clandestine operations have expanded considerably since then and they now work closely on anti-terrorist operations with the Army's Special Air Service. Here in the 1960s, Mne J Michie (right) and his team prepare for a night operation from one of HM Submarines, a technique which has been perfected over many years.

LVTs manned by the Royal Armoured Corps, with later waves in LCAs. After the naval bombardment, coupled with close air support from the Fleet Air Arm, there was little opposition. 45 Commando (Lt-Col N H Tailyour) then carried out the first-ever unit helicopter assault. They landed in relays on the secure beach near the De Lesseps statue from helicopters of 845 Naval Air Squadron (Whirlwinds) in HMS *Theseus* and the joint Helicopter Experimental Unit (Whirlwinds and Sycamores) in HMS *Ocean*. Twenty-two helicopters in all carried the complete force of 400 officers and men, plus 23 tons of stores, to Casino Pier, on the left of the bridgehead, in 91 minutes. A remarkable feat considering that the smaller helicopters only held three passengers.

Once British tanks from 6 Royal Tank Regiment had rolled ashore, the advance continued southward into the town, with 42 Commando sending 'A' & 'B' Troops, re-embarked in their LVTs, on a hazardous dash through the town to secure the power station and cold storage depot on the southern limits of Port Said, while 40 Commando slowly and systematically cleared the port area adjacent to the canal, which included most of the municipal and government buildings. Although there was comparatively little opposition, what there was was tough and determined. A particularly bloody fight ensued through the Canal Company's HQ, the port Police Station and Navy House, where two 40 Commando officers, Lt P W McCarthy and Lieut E A Ufton were killed leading attacks through the labyrinth of customs sheds and offices, and many marines were wounded. Meanwhile, 45 Commando had the unenviable job of sealing off the Arab town and preventing armed mobs, whipped into a frenzy of agitators, from intervening in the fighting.

Further objectives for the following day were defined and a continuation of the advance down the length of the Canal planned. But behind the scenes enormous political pressures had been applied and at midnight, much to the disbelief of those involved in the fighting, came the announcement of a ceasefire. By this time advance troops had reached El Cap fifteen miles down the Canal. That night ended with mobs looting and pillaging in the customary Egyptian way. The Commandos were forced to disperse one stone-throwing crowd of 500, who were rioting over sacks of flour looted from an Egyptian army depot. For several days the marines patrolled the town and on one occasion a search of patients in a hospital revealed weapons hidden between the mattresses and bed-springs. Eventually 3 Commando Brigade was replaced by army units, and later a United Nations force took over. It was ironic that some of 42 Commando were not allowed to fraternize with Norwegian troops of this force with whom they had been exercising in Norway only four months before! Thus ended the 'Suez Incident' - all over in a few days, but for which it had taken months to prepare. It was an action which, thirty years on, still arouses powerful emotions at home.

The slowness of the seaborne assault force, which took a week to sail from Malta to Port Said, threw into disfavour conventional landing craft and methods perfected in 1944 and 1945. The helicopter assault by 45 Commando from the light fleet carriers showed clearly the way forward for the future. From this experience, and with the backing of the First Sea Lord, Admiral Lord Louis Mountbatten, came a new concept which was to carry the Corps through the 1960s and 70s. This concept was the 'Commando Carrier', which applied until a fresh retrenchment of conventional naval forces began anew with the arrival in 1981 of John Knott and his retrograde ideas at the Ministry of Defence.

3 Commando Brigade with 40 and 45 Commandos, returned to Malta, while 42 Commando was once more reduced to a reserve at Bickleigh. They did however re-activate in the summer of 1958 for the Lebanon crisis, but only got as far as exercising in Libya before returning to UK. The Cyprus situation, however, was far from resolved. The two units in Malta found themselves alternating there over the next two years, operating mostly in the mountainous Troodos region against the even more determined EOKA terrorists, and part of 45 Commando formed a 'Heliforce' for a period in 1958. After the successful use of helicopters during the Port Said landings, 45 Commando carried out many trials and operation in Cyprus with their Commanding Officer, Lt-Col Jack Richards, leading a special force of 'sky-cavalry' provided by RN 728 (Commando) Flight of four Whirlwinds. The unit's final tour in the trouble-stricken island finished on 15 December 1958 and they spent a peaceful Christmas in Malta with their families. 40 Commando did one further tour there in the Spring of 1959.

When the aircraft-carrier *Bulwark* was taken in hand for conversion, rejoining the fleet in 1960, she had all her normal equipment, aircraft and launching gear stripped from her. Her hangar decks now accommodated a helicopter squadron, trucks, land-rovers and other vehicles, assault craft and full stores to equip a 600-strong Royal Marine Commando. The first unit to embark in her was 42 Commando and together they sailed from Plymouth for the Far East in March 1960, exercising in the Mediterranean on the way. In 1961, Lieut Roger Learoyd became the first RM Officer to join an operational helicopter squadron as a fully-fledged pilot. Meanwhile, her sister ship *Albion* was similarly converted, joining the fleet in April 1962: other conversions were to follow. Following American lines two special assault ships *Fearless* and *Intrepid*, were laid down in 1962 and others of a slightly different design were planned. Although close-range armament was retained these ships lacked conventional weapons and relied on other warships for their main protection.

But more trouble was not far away and in April 1960, 45 Commando were once again on the move, this time to Aden, where they became permanently based. Here the troubles were Yemeni-inspired,

HMS Bulwark, *the first Commando Ship and known affectionately as the 'Rusty B', looking very shipshape as she prepares to enter harbour in the Far East in 1967. The ship's company are fallen in on the starboard side with 42 Commando lining the port side. Forward are the guns of the Commando Light Battery Royal Artillery, while the Commando's flight of Sioux helicopters and the Naval Air Commando Squadron's Wessex are ranged amidships, with the Commando transport aft.*

Rock climbing has always been an integral part of Commando training and the rugged west country coastline has proved an ideal testing ground. The Cliff Climbing Wing (later Cliff Assault Wing) was originally based in St Ives, Cornwall, moved to Bickleigh in the mid-1950s, but retained their Cornish training areas. Here an Army exchange SNCO leads trainees high above the beaches on a rock run that helps to develop personal skills and confidence.

reflecting general unrest in the Arab world. For just over seven years the Commando, with a trickle drafting system giving an unaccompanied one year's tour to thousands of Royal Marines, fought with great distinction in the desert wastes and barren mountains of the Aden Protectorate at Dhala.

Thousands of post-World War II marines will remember their often very wet baptism in amphibious landings, which took place at Jennycliffe to the east of Plymouth Sound. Here recruits land from a dory in 1960 as part of their Commando course.

40 Commando were based on Malta, and to provide a much needed reserve for the carriers, 41 Commando was re-formed at Bickleigh in March 1960, and this mini-expansion continued with the

re-forming of 43 Commando at Plymouth on 5 September 1961. In addition, the strength of each Commando was increased to 687 officers and men. For what became known as 'brush fire' wars, small-scale insurrections and the like which might be nipped in the bud before they became major wars, the Commando Ships enabled a quick response around the globe, especially in threatened newly-independent Commonwealth nations where British influence and assistance were still valued. Such a role was highly fitting for the Royal Marines. National Service, or conscription, had now ended and once more the Corps was an all-volunteer force of the very highest standard.

In June 1961 Iraq mobilised and threatened to take over the rich oil state of Kuwait. The ruler of Kuwait was a good friend of Britain and requested help. By a stroke of luck *Bulwark*, with 42 Commando embarked, was off Karachi only two days steaming from the Persian Gulf. On arrival, the unit was flown ashore to dig in along the border between the two

countries on Mutla Ridge while the Amphibious Warfare Squadron, also on hand, started to unload heavy equipment. 45 Commando were air-lifted out of Aden post-haste and the small Royal Marines detachment from HMS *Loch Alvie* was also landed. The presence of these marines, together with their supporting squadron of tanks and other army units, was more than enough to persuade Iraq to change its attitude.

During the early months of 1961 Headquarters 3 Commando Brigade moved from Malta to occupy the former Royal Naval Air Station at Sembawang in Singapore, later to become the Fleet Amphibious Forces Base, Far East. 40 Commando followed them in May 1962, and were initially based in Johore. During the next few years a large number of exercises were carried out, ranging from Aden to Hong Kong, Australia to Sarawak and also with the US Marines in the Philippines. In addition there was an internal security role to play in Singapore. For the first time since the war a major re-organisation within the commandos took place. The five Troops were replaced by three Rifle Companies and Support and HQ Companies. These companies were lettered throughout the alphabet for easy identification. Whereas Troops had been lettered similarly to the turrets of a capital ship

When the Ruler of Kuwait asked for British military assistance against a possible invasion by Iraqi forces in June 1961, 42 Commando, already embarked in HMS Bulwark *were exercising two days sailing away. They landed and took up positions on Mutla Ridge, supported by army reinforcements, which were enough to deter the Iraqis, showing the value of the quick reaction of the commando ship concept.*

('A', 'B', 'X', and 'Y', etc), the new companies were 40 Commando ('A','B','C'), 41 ('E','F','G'), 42 ('K','L','M'), 43 ('O','P','Q') and 45 ('X','Y','Z').

In keeping with the traditional role of the Royal Marines was a call on their amphibious expertise and military skills to deal with a revolt which broke out in Brunei, the British Sultanate of north-western Borneo. This took place in December 1962, and after a company of Gurkhas had found the situation more than they could cope with, 42 Commando were flown out from Singapore in record time, landing at Labuan and assembling in Brunei Town. Here they discovered that rebels had taken as hostage the British Resident and other Europeans, including two women, and were holding them 12 miles up river at Limbang settlement.

Improvisation was ever the way of the Royal

Casualties returning to Brunei Town after the successful raid on Limbang on 12 December 1962, when L Company, 42 Commando, released about a dozen British hostages, including two women, and one American unharmed. The Company lost 6 men killed, whilst police casualties were five dead. This was one of the Z Craft commandeered and refitted quickly for the raid.

Marines and, evoking memories of Gallipoli in 1915 and Tobruk in 1942, two old 'Z' lighters that belonged to the Brunei government were discovered at the end of the waterfront. The two coastal minesweepers, HMS *Fiskerton* and *Chawton* provided crews for these vessels, and 'L' Company embarked to deal with the situation. After a perilous night journey through narrow rivers covered in mangrove swamp they arrived off Limbang at dawn on 12 December, and in a daring frontal assault against small-arms fire, took the rebel HQ and mopped up the remainder, the marines losing five killed and five wounded in the process. The rebels broke into the jungle but more importantly all the hostages, who were about to be executed, were released unharmed.

When the revolt erupted, 40 Commando were exercising aboard *Albion* off Mombasa and were hurriedly rushed to the scene, landing with Brigade HQ at Kuching, the capital of Sarawak, on 14 December. They were quickly moved north and were soon operating in the jungle having been landed by *Albion's* naval helicopters. Within a week the rebellion had been quashed and the remaining rebels driven into the jungle. 42 Commando earned an MC and two MMs for their brisk and highly efficient action, which undoubtedly showed the rebels that the British were still capable of quick and decisive intervention.

For the next three years, Brigade HQ and the two Commandos took their turns with army battalion to quell the insurrection which had not only been driven across the Indonesian border, but was now backed by the Indonesian regular forces. Independence within the newly-formed Malaysia was granted to Sarawak and North Borneo, renamed Sabah, on 16 September 1963, but communist inspired unrest was not far below the surface. The designs of the Indonesian government, under President Soekarno, against the Malaysian Federation led to a period of what was simply known as 'Confrontation'. Once more infiltration by Russian trained and armed rebels through the jungles began, while the Indonesian Air Force and Navy were equipped with Soviet-built warships and aircraft to support them. This threat made it necessary to base powerful naval forces on Singapore, and there was always one Commando carrier at hand for emergencies. They sent their helicopters into jungle outposts, sometimes more than 100 miles from the sea, until the RAF was able to build up their own helicopter force. Like all jungle campaigns, it was a war of stealth and ambush. Small marine patrols ventured hundreds of miles inland, often being taken there via the rivers aboard the small coastal minesweepers. A Special Boats Section played a highly important role in conducting clandestine operations both deep into enemy territory and along the coast, bringing a wealth of intelligence to help the security forces.

As usual the British 'defenders', based at Kuching and Tawah (in Sabah), were hamstrung by nervous governments, and decisions taken back home. Another major problem was the ill-defined border, an imaginary line drawn across impenetrable jungle, which was strictly out-of-bounds to the marines but, of course, not to the Indonesians who crossed and re-crossed it to carry out their raids. Nonetheless, Britain stood up to this latest tin-pot dictator as she had done with Napoleon, Hitler and Mussolini. On one occasion, a small naval task force sailed through the Lombok Strait, international waters barred by Soekarno, ready for action, but nothing happened.

On 31 August 1965, a large patrol from 'B' Company, 40 Commando, fought a brisk action when they bumped an Indonesian platoon which had crossed the border. The marines quickly accounted for six of them dead and wounded four more before the remainder fled, at the cost of but one casualty to themselves. This action reinforced their superiority at such jungle craft and eventually this undeclared war petered out. Many such resolute actions were fought by the Royal Marines during those three years and 36 gallantry awards were won.

It was during this time that unit light helicopter flights were first formed, the first by 42 Commando in 1965, their three Sioux helicopters being piloted by RM Officers. Subsequently on 12 August 1968 3 Commando Brigade Air Squadron was formed with a flight to support each Commando, being known by a unit designation - 40 Commando had Dieppe Flight, 41 Salerno, 42 Kangaw and 45 Montforterbeek Flight.

Meanwhile further west across the Indian Ocean, trouble brewed in a number of places. In January 1964, Africa was the scene of another impressive demonstration, when President Nyerere of Tanganyika had to call for British help to quell an armed insurrection by rebellious troops of the East African

A typical scene during the Commando ship era. When the Brunei revolt broke out in December 1962, HMS Albion *was exercising off Mombasa with 40 Commando embarked. They sped across the Indian Ocean and had no more than a brief resupply and refuelling stop in Singapore. Four days later many of the commandos were dropped into the jungles of the northern district of Sarawak by RN Wessex helicopters.*

During the 1960s Royal Marines Commandos were often embarked for several weeks. Space for training was mostly limited to the flight deck which had to be shared with the RN Helicopter Squadrons. On passage to Hong Kong in HMS Bulwark *in 1967, marines of 40 Commando practised Internal Security drills.*

The military use of the hovercraft has never been fully exploited but one of the first 'pilots' of the Inter-Services Experimental Hovercraft unit was Capt Stuart Syrad. The craft made light work of jumping the Borneo rapids, but its excessive noise made it of doubtful military value. This SRN 4 negotiates the rapids of the Rajang river in 1965.

army. HMS *Centaur*, a conventional aircraft carrier, acted in the commando ship role by hastily embarking 45 Commando at Aden. She arrived off Dar-es-Salaam on the morning of 25 January and flew the marines ashore. They quickly established themselves at the Colito Barracks, the hotbed of the revolt, and with Royal Navy fighter/bombers overhead, they soon subdued the native troops. 41 Commando was flown in from the UK to relieve 45, but the revolt was snuffed out in a few days.

45 Commando returned to Aden where the well-rehearsed role of holding down a rebellious colony until it was granted independence was played out once more. The former coaling station for the fleet and strategic stopping place on the route to India still had a significance in the defence of the Middle East oil-fields, even though both Suez and India had been given up. The usual pattern of the 1960s emerged, when communist-armed and inspired terrorists began infiltrating from neighbouring Yemen,

The Massed Bands of the Royal Marines Beat Retreat on Horse Guards Parade, London, in honour of the birthday of the Captain General, HRH The Duke of Edinburgh in 1967. This ceremony is normally performed every third year and this was the last appearance of Lt-Col Sir Vivian Dunn, who was the first Principal Director of Music Royal Marines, and had held this appointment since 1953.

On 2 December 1964, as part of the Corps Tercentenary Year, a British Railways Type H Diesel Locomotive, No D 70, 'The Royal Marines' at St Pancras Station, London. The Commandant General, General Sir Malcolm Cartwright-Taylor, unveiled it and, after presenting the crew with inscribed tankards, was allowed to drive it. The engine ran daily services on the London Midland region, mostly pulling goods trains and 'went to pension' after 22 years in 1986.

42 Commando have paraded a pipe band from time to time since their service in the Far East in World War II. Their dress has varied from standard khaki drill in the late 40s to these tunics and kilts, made from service blankets. In 1968 they played at the opening of the Ulu Community Centre in Sembawang, Singapore.

Whilst 3 Commando Brigade was stationed in Singapore from 1960-1972, opportunities were taken to carry out commando ship exercises with Commonwealth forces. A 'stick' of marines of 42 Commando land from a Wessex helicopter from HMS Albion *in the Shoalwater Bay area of North Queensland in Australia in October 1968 during exercise* Coral Sands. *This was the largest Pacific amphibious exercise since World War II with 3 aircraft carriers, 2 Landing Ships Logistic and many smaller warships.*

The chaplain of 45 Commando, the Rev Ray Roberts (later to become Chaplain of the Fleet) conducts a service in the Radfan region of the Aden Protectorate in 1963. Troops always had to be alert in this dangerous, mountainous area, and here can be seen a signaller and a marine with a GPMG.

Whilst operating in the Far East in the 1960s close liaison was maintained with locally enlisted forces. Cpl S Carlisle at work in the Radio Operations Room on Sebatik Island in Sabah (North Borneo) with a signalman from the Malay Federation Signals in September 1964.

The Life Colonel Commandant pays his first visit to Lympstone on 28 October 1966 and is shown the 'water-tunnel' on the endurance course by Capt B Le Mesurier. From left: Maj-Gen F N Grant (MGRM Plymouth), Lt-Col C A Watkins (Chief Instructor), Col L G Marsh (Comdt), and in background Lt C R J Goode (AdC to MGRM), Major M A H Marston (Chief Trg Offr) and Capt D A Oakley (AdC to Earl Mountbatten).

carrying out murderous and indiscriminate attacks on British servicemen and civilians.

To this bleak and dusty corner of the Arabian peninsular came the Royal Marines of 45 Commando, evoking memories of earlier campaigns in the eastern Sudan. The mountainous Radfan district of the Aden hinterland became the backdrop for the Commando to carry out dozens of patrols and searches, which from time to time ended in brief and bloody fire-fights with the rebels. Once more they were the last to leave the Radfan when the withdrawal of British troops from the colony was ordered. They took their turn in the infamous Crater district of Aden when their individual restraint was proof of their excellent training and discipline. When the withdrawal eventually came, they were joined by 42 Commando from Singapore who covered the departure of 45 Commando and were therefore the last unit to leave although the final foot on Aden soil was probably that of the bowman of a Royal Marines landing craft that evacuated the last wave on 29 November 1967: they returned to England and to Stonehouse Barracks at Plymouth.

On the ceremonial side, in 1964 the Corps had celebrated the Tercentenary of its formation as the Lord High Admiral's Regiment. Her Majesty The Queen reviewed the Royal Marines in the grounds of Buckingham Palace on a perfect summer's day on 23 July and that evening dined with the officers of the Corps at Greenwich. It was here, over dinner, that the Chief of the Defence Staff, Admiral of the Fleet the Earl Mountbatten of Burma suggested that she might like to commemorate the occasion, as her father had done when he dined with the Corps in 1948, by granting the Royal Marines the privilege of drinking the Loyal Toast seated in all their messes ashore and afloat. This privilege was first exercised by the senior non-commissioned Officers of the Corps when they dined the Captain General at Portsmouth Guildhall on 30 July. On the last day of Tercentenary Year, 27 October 1965, Earl Mountbatten of Burma was welcomed into the Corps as Life Colonel Commandant by his nephew, the Captain General, at an impressive parade at Eastney. It was Earl Mountbatten who had done so much for the Corps, particularly since his days as Chief of Combined Operations in 1942. It was at this time that Lovat uniform was introduced as an alternative to Blues and to replace battledress, and so from 1 April 1964 it was worn for minor ceremonial, by duty-men and for walking out.

Thus, although much reduced in numbers, the Corps entered the 1970s feeling that they had come through their worst period and were in good shape to look ahead to a worthwhile and promising continuation of their new role and the beginning of another 300 years of service.

HM The Queen and HRH The Duke of Edinburgh watch Beat Retreat before dining with the officers of the Corps at the Royal Naval College, Greenwich on 23 July 1964 to mark the Corps Tercentenary Year. Accompanying Her Majesty are the Commandant General (General Sir Malcolm Cartwright-Taylor), the Minister of Defence RN (Earl Jellicoe) and the Admiral President of the College (Rear-Admiral I S McGeoch). In rear is the Representative Colonel Commandant (General Sir Campbell Hardy).

Royal Marines musicians are taught to play string instruments for orchestral work, in addition to playing in the military band. In the 1940s and 1950s Junior Musicians were recruited from the age of 14 1/2 and were given full time education as well as their musical and military training.

A patrol of 7 Troop, Z Company of 45 Commando in the Radfan region in 1963. This photograph has often been mistaken for one of the Parachute Regiment, but the Marines are readily identifiable as Cpl D Dorling, Mnes Goffett, Gibson, Sykes and Allen. This shows the rugged, mountainous and inhospitable countryside over which the Commando fought for more than three years.

On 30 July 1964, as part of the Tercentenary celebrations, HRH the Duke of Edinburgh dined with 186 Senior NCOs of the Corps at the Guildhall, Portsmouth. On the top table can be seen from left RSM D Dean, Maj-Gen R D Houghton, RSM G A Bray, General Sir Campbell Hardy, RSM W Chisholm (Senior RSM of the Corps), General Sir Malcolm Cartwright-Taylor, Maj-Gen N H Tailyour. This was the first occasion since the new privilege of drinking the Loyal Toast seated was granted, that it was observed.

The lying in state of Sir Winston Churchill in Westminster Hall, London, from 25-29 January 1965. This was the first time that the Royal Marines had the honour of providing one of three Royal Navy watches. Seventeen officers took part in the vigil, including the co-author. They carried out two watches of six hours, each officer being on duty for 20 minutes at a time with a 40 minutes stand-off.

CHAPTER 11

1971-1988
COMMITMENT TO NATO

'We didn't mind dying for Queen and country, but we didn't contemplate dying for politicians!'

Major General Julian Thompson.
ITV interview on The Falklands War. April 1987.

After losing 43 Commando in the autumn of 1968, the strength of the Corps was reduced to 633 officers and 7,515 NCOs and men, about 800 of whom were in the Royal Marines Band Service by 1970. There were also 118 officers and 851 men in the Royal Marines Reserve. Severe reductions threatened the very existence of the Royal Marines once more. The continued withdrawal from Great Britain's world role was reflected in the pull-out from the Far East in the early 1970s. Singapore was abandoned and Brigade HQ, 40 and 42 Commandos had all returned home by 1972. They were initially concentrated in the West Country. Only 45 Commando, which became based in Arbroath in 1971, had a specialist role, acting on NATO's northern flank responsible for north Norway adjacent to the Soviet Murmansk complex, the largest military and naval base in the history of the world. The Corps had been seeking a new role, and here they were able to specialise in arctic warfare operations, becoming ski troops, equipped with SnoTrack vehicles and working closely with the Royal Netherlands Marine Corps. The Dutch Marines formed 'Whisky' Company in 1973 under operational command of 45 Commando and they proudly wear the unit's red lanyard. Together they participated in numerous mountain warfare exercises throughout the winters of the early 70s, including quick response landings in remote Norwegian fjords. This new role reflected what 42 Commando and the Commando School had been doing during the 1950s, when they carried out snow warfare training at Aviemore in Scotland and held two exercises around Bardufoss airfield in Norway.

With the concentration of the home bases in the Plymouth area, the old hutted camp at Bickleigh was

Not only the Special Boats Section exercise regularly with submarines, but members of 539 Assault Squadron, formed in 1983, and Z Company of 45 Commando practice launch and recovery drills with HMS Oracle.

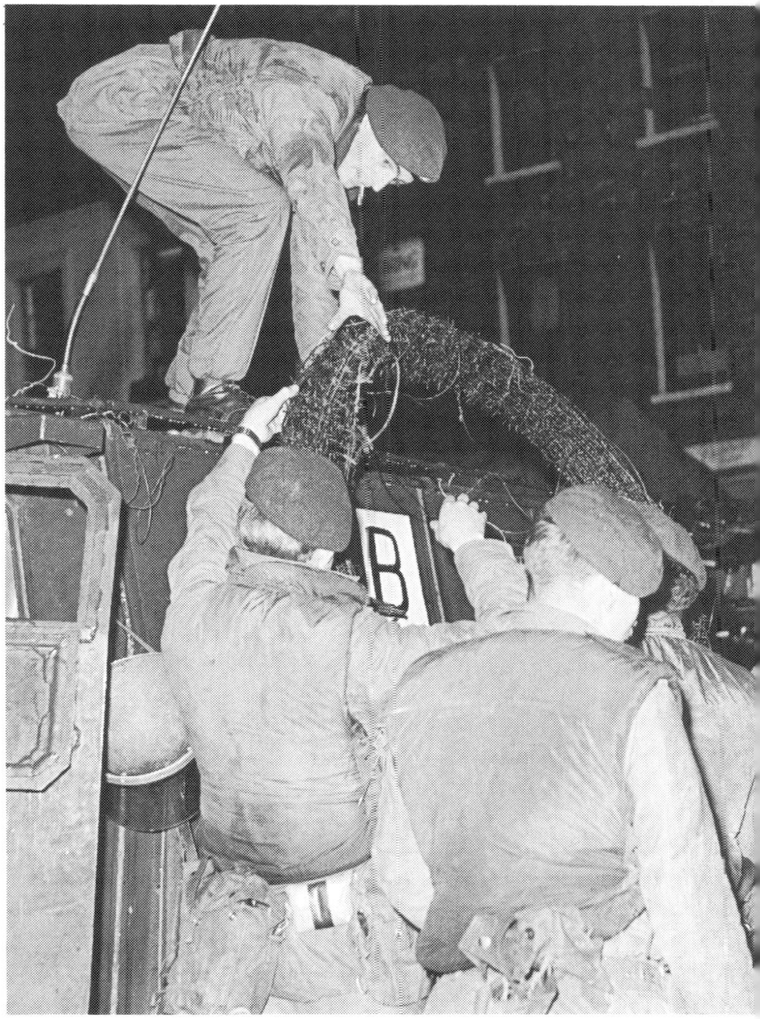

The Shankhill Road rioting of September 1970 saw some of the worst violence at the beginning of the renewed Ulster problem. 41 Commando were in the thick of it, having relieved 45 Commando on 1 September, only 9 months after their first tour. Marines are seen erecting barricades in the Shankhill District, a common part of Belfast night life. On 30 October, 41 Commando were in the thick of the action again in the Crumlin Road riots.

re-built as a barracks, and the Amphibious Training Unit at Poole was also modernised by 1971. Poole town celebrated this new status by granting the Freedom of the town to the Corps on 5 September 1973, while CTCRM, the name had changed from ITCRM on 24 August 1970, at Lympstone continued in its enlarged role with new buildings springing up regularly to meet modern conditions. In contrast, Eastney was due for closure in 1973 and held the last Corps Remembrance Parade that year. The function of Technical Training which included drivers and tradesmen, previously the prerogative of Fort Cumberland, were moved to Poole, while the Signal Training Wing re-located at Lympstone. However

Trouble flared in Northern Ireland in 1969 when 41 Commando were sent to the Province. The following year 45 Commando followed as reinforcement for the General Election and Orange Day celebrations. 10 Troop of Support Company help to keep Catholics and Protestants apart during some particularly violent rioting on Black Friday, 26 June.

Eastney, did see a resurgence when the new Corps Museum was opened by HRH The Duke of Edinburgh in the old Officers Mess building on 1 August 1975, and this has become a premier service museum in the country with its splendid setting and outstanding renovations that are the envy the world over, a fitting show-case for an illustrious Corps. In addition, Eastney retained its prime role as the Headquarters of Training Group.

On the ceremonial side, Royal Marines Concerts began on an annual basis at the Royal Albert Hall in 1973 and continue to this day, now being commemorated as 'Mountbatten Festivals of Music'. So popular are they that the one evening originally allocated has now stretched to three nights, and thousands of pounds have been raised for the Malcolm Sargent Cancer Fund for Children and Royal Marines Charities. Another affiliation was that with the church of St Lawrence Jewry at the Guildhall on St George's Day 1974, cementing relationships with the City of London which had begun with the Trained Bands 310 years earlier.

Although Royal Marines detachments at sea had been confined to only a few ships including commando carriers, the main commitment was reduced to small detachments of ten men in frigates. HMS *Phoebe* had carried the last officer-commanded 22-man detachment, but by the end of the decade sergeants were mostly the senior Royal Marines at sea.

Trouble has never been far from our shores and the Corps were soon committed to Ulster. In 1969, 41 Commando, as *Spearhead* battalion were sent to Belfast, where rioting had broken out. At the time there was little modern-day internal security equipment, such as batons and shields, and the Marines faced their tormentors unprotected. The following year, the Commandos started the first of many tours to the troubled province, which continue to the present day. The explosive situation is not so far removed from that which faced the marines in 1920 and later in 1957. It was, and is, a religious war of separation which is incomprehensible to most thinking marines, and it was the first time that many marines found themselves facing their own kin-folk with bullets in their barrels. Four months tours by Commandos, taking their turns with army battalions, have now become an accepted, but nevertheless disagreeable, part of modern service life.

Meanwhile Commandos still exercised in many other parts of the world. 40 Commando returned to Hong Kong in 1970, their last tour from Singapore before leaving the island for good on 30 September 1971. Earlier a large company group, along with elements of HQ 3 Commando Brigade and 847 Naval Air Squadron did valuable work in the flood disaster of East Pakistan (Bangladesh) in November 1970. It is in such disasters as these that Royal Marines, with their inherent amphibious expertise and individuality, can play such an important role. 42 Commando also paid a farewell visit to Limbang and Sarawak the same year, and then served aboard *Albion* in the Mediterranean. 3 Commando Brigade held their final parade at HMS *Simbang* on 18 March 1971 and returned home after spending 28 years consecutively overseas.

But for the most part, the Corps, like the nation, had turned their sights inwards. The terrorist bomb had burnt a bloody path of retreat all the way from Ulster to Oxford Street. After thirty years of gradual relinquishment of our overseas commitments, the Commando Brigade had come home to roost. Riots sparked off from Tottenham to Trinidad, Harmsworth to Guyana. Royal Marines found themselves in the centre of many of these outrages, and young marines prepared for the same booby-traps, bombs and bullets that had faced their fathers in Aden, Palestine and Cyprus in the past.

British ties with Norway have always been strong and here King Olav of Norway inspects a Royal Guard from the detachment of HMS Hermes *in 1973. One of the hazards of arms drill on the quarterdeck of an HM Ship was the second movement of the 'Present Arms' when the awning was spread. Many a naval sailmaker has had to repair the subsequent slits!*

HRH Crown Prince Harald of Norway was appointed Honorary Colonel of the Royal Marines on 5 May 1981. This was the first such honorary appointment in the Corps of a member of a foreign royal household and symbolised the close relationship developed over the past 50 years between the Royal Marines and Norway. In 1982 he visited HMS Victory *in Portsmouth on which there was a permanent RM detachment who acted as guides for visitors.*

Wearing flak jackets, the Band of Commando Forces, under Capt Terry Freestone, give a military band concert in Belfast City Centre in 1972. Visits of bands to trouble areas, as had previously been carried out in Hong Kong, Aden and Borneo, are tremendous moral boosters to both marines and civilians alike.

On 31 July 1972 over 20,000 British troops flooded Northern Ireland and both 40 and 42 Commandos were involved with Operation *Motorman*, when bulldozers and other heavy equipment was brought ashore in the Royal Marines manned LCMs of HMS *Fearless*. The crews of landing craft and small boats have played an integral part in counter-terrorist operations in Ulster ever since. Like so many of these explosive internal security operations in the past, success has been measured, not in kills and captures, but in peace and the restoration of a normal way of life to the inhabitants. By the end of 1980, Royal Marines Commandos had carried out 22 tours of the province with 11 officers and men being killed and more than 80 wounded. The 1980s have been quieter and the IRA's methods have become more sophisticated, requiring quick reactions, coupled with diligence, patience and a sense of humour.

One of the deficiencies in the Corps' armoury had always been a lack of integral support, relying almost entirely on the army. During the Borneo confrontation it had become abundantly clear that gunner, sapper and logistic support was needed and various *ad hoc* steps were taken. The first major development was the formation of 29 Commando Light Regiment Royal Artillery in 1961 with 145 (Maiwand) Battery firing their first shot in anger in Borneo on 23 December 1962. They were joined two years later by the conversion of the Amphibious Observation Battery at Poole who became 95 Commando Light Regiment. On return from the Far East, gunner batteries were allocated to Commandos 145 (Maiwand) at Arbroath, 8 (Alma) and 79 (Kirkee) in Plymouth, while 148 (Meiktila) Forward Observation Battery served at Poole. They also took their turns serving as infantry in Northern Ireland.

Between 1968 and 1971, 59 Field Squadron Royal Engineers worked closely with 3 Commando Brigade, and in April 1971 re-formed as 59 Independent Commando Squadron RE. Every gunner and sapper serving with these units has to pass the Commando Course to earn the coveted green beret. Joining them in this arduous training came RAOC, RCT and REME personnel who make up nearly fifty per cent of the Commando Logistic Regiment Royal Marines, which held its 'commissioning' parade on 11 January 1972. Included in the regiment are Royal Navy personnel of the Commando Medical Squadron, doctors,

Cpl Dave Winn pictured in an off duty moment during the Calcutta Flood relief in October 1978. With Sgt Peter Biggs and Cpl Dave Oates from 1st Raiding Squadron, he trained 200 Indians as coxswains in 5 days in the use of inflatable raiding craft. Sgt Biggs was awarded the BEM and the two JNCOs gained Commandant General's commendations.

dentists and medics all of whom have also qualified for their green berets. The regiment, about 400 strong is commanded alternately by a Royal Marine officer and an Army officer.

Great Britain still held many interests in the Mediterranean and 41 Commando were sent as garrison troops to Malta on 3 September 1971. They exercised as far afield as Turkey and Puerto Rico during the ensuing years, making an important contribution to NATO's southern flank. The unit was temporarily 'deported' from Malta for political reasons during the summer of 1972 and spent the winter of 1974/75 in Cyprus as part of UNFICYP. Here they laid aside their green berets and temporarily adopted the light blue ones of the United Nations Forces. Periodically other Commandos joined them in the Mediterranean, working from the commando ships in exercises involving Americans, French, Italians and Turks. These were all part of exercises that continued through the 1970s under the names of *Dawn Patrol, Deep Furrow,* and *Display Determination.*

Mnes Chris Neeve and Chris Johnson of E Company, 41 Commando, wearing the light blue beret of the United Nations peacekeeping force in Cyprus, talk with the Life Guards crew of a Ferret armoured car in Arlona in 1979.

Yet more cuts in Corps strength necessitated 41 Commando being reduced to one large operational company in April 1977, Salerno Company, who continued to provide the garrison for Malta, but even they left in March 1979. However the Unit took up residence once more re-formed, at Deal and, apart from exercising in many parts of the world and taking their turn in Northern Ireland, had the honour of representing the Royal Marines when in November 1978 the Corps were asked to perform London Duties

Members of the Special Boats Section diving in the Arctic where the water temperature is –10°C. Pictured at Harstad, this marine wears RN issue Aquarius breathing apparatus before diving from a rigid inflatable craft.

'If you are only three feet high, inspections can cause your view to be somewhat restricted!' Nigel Wheaton was a nine-year-old member of the RM Volunteer Boys Corps at Plymouth in 1979.

The Royal Marines, represented by 41 Commando, were invited to perform Public Duties in London for the second time in their history from 4-30 November 1978. For the first time a guard was mounted at the Tower of London in addition to Buckingham Palace and St James' Palace.

for the second time in their history. They provided guards not only for Buckingham Palace and St James's' Palace but also an additional guard at the Tower of London for the first time.

At sea, Royal Marines detachments, although small, were often involved in policing duties, one of these being the 'Cod' war off Iceland during 1976/77. HMS *Protector*, the ice patrol ship, carried a detachment from its earliest days in the 1950s, patrolling in the Falklands Islands area of the South Atlantic. In fact, in 1953, after HMS *Superb's* Royal Marines had been landed on the islands, a platoon was sent out in HMS *St Austell Bay* to relieve them. Naval Party 8901 of about 25 Royal Marines was formed as the permanent garrison stationed there in 1965, the detachment serving a one-year tour. In September 1966, 20 Argentinian 'nationalists' hijacked an airliner and landed at Port Stanley to try to capture the islands, but the situation was soon brought under control as Royal Marines surrounded the plane and the whole incident passed off quietly without a shot being fired. In recognition of their special and long association, the Falkland Islands honoured the Corps by granting them the Freedom of Stanley on 8 December 1976.

At home, the commitment to the Northern Flank had seen an advance party from 45 Commando setting up a base at Elvergardsmoen, near Narvik, on 2 January 1970 and these were the forerunners of a

Moody Brook Camp in the Falklands, the home of NP 8901 before the Argentine invasion of 1982. Life was comparatively peaceful, patrolling being confined to visiting the kelpers in the homesteads and watching the abundant wildlife. On the domestic front Mne Philip feeds some 'new recruits', which helped to supplement the rations.

Colonel Sir Steuart Pringle received the presentation sword from the acting Governor of the Falkland Islands, Mr Gordon Slater, when the Freedom of Stanley was conferred on the Royal Marines on 8 December 1976, the anniversary of the Battle of the Falklands in 1914, in which 653 officers and men had fought. The Corps connections with the islands go back to 1766, when Great Britain established its first settlement on Saunders Island and the Marines built a Blockhouse there. The Royal Marines were present when the Settlement moved to Stanley in 1842 and have been there continuously since 1953, either as the detachments of HMS Superb, St Austell Bay *and* Protector *or as Naval Party 8901 until soon after the Falklands War of 1982.*

continuing series of winter exercises there. The Corps had found itself a new and exciting role and the old Cliff Assault Wing, changing its name and functions twice in the process, settled in Arbroath in 1970 as the Mountain and Arctic Warfare Cadre and moved to Plymouth in 1982. In the first winter 242 Marines savoured the delights of 10-man tent groups, pulk-pulling and military skiing, most for the first time. From those humble beginnings the Corps are now the acknowledged experts in arctic warfare and survival, and every marine is now likely to spend at least one winter in northern Norway, exercising and operating with the Norwegian army and other NATO allies.

Memories of the coal strike of 1921 were rekindled by the 'Winter of Discontent' public services strikes in 1977/78, and once more the specialist services of the Corps were called upon to help out in a national crisis, running fire appliances and the like for several months.

But the diminishing policing role was reflected in the gradual run-down of the commando ships. The two old light-fleet carriers soldiered on long past their allocated life-span. *Albion* being paid off in 1973, but the incredible 'Rusty-B', the ancient *Bulwark*, kept panting across the world's oceans until 1981 when she was finally laid to rest. She had been laid down in 1944. Plans to replace them with another ship *Hermes* dragged on due to lack of funds, but she was eventually de-commissioned as a fixed-wing carrier and converted to carry out the dual roles of anti-submarine warfare carrier and commando ship. It was not a happy 'marriage'. She re-joined the fleet in

Colours have to be replaced from time to time when they become too worn to carry. 40 Commando were presented with new Colours by HRH The Duke of Edinburgh at Plymouth Hoe on 1 July 1976, and the old Colours were laid up in the Royal Marines Museum at Eastney on 27 January 1977, where they are now on display. The Queen's Colour (Lt A I B Troup) is on the left and the Regimental Colour (Lt R M Hill) on the right with the second-in-command Major S R C Bemrose in the centre. In the rear are the last Colours of RM Barracks Eastney which were laid up in 1973.

1973 and served for more than another decade. Later the advent of the V/STOL fighter aircraft saw the addition of a take-off ski-lift 'hump' forward for Harriers which was added to her other roles. She was no larger than her predecessors, but carried these extra functions until she was also paid off in 1981, in readiness for scrapping or sale to Argentina. However that nation was to grant her a reprieve as history subsequently showed, and she earned her greatest glory in the twilight of her life. The two amphibious assault ships *Fearless* and *Intrepid* which had come into service during the late 1960s were still available on a part time basis for exercises with the Commandos.

Another new role for the Corps came with the development of British interests in oil-fields in the North Sea. It was on 20 July 1977 that 'L' Company of 42 Commando was first deployed in what today is described as an 'Oilsafe' role. Their task was evaluate the methods whereby they could protect an oil-rig, often hundreds of miles from the shores of Britain, against terrorist attack. The Royal Navy's need to defend its own installations in the UK became more of a reality as sabotage threats grew. And so in May 1980 Comacchio Company RM, taking on many of 43 Commando's old traditions (and its red and old gold colour lanyard), was formed and based in Arbroath. Initially it had a strength of about 300 all ranks, but with an increase in its commitments, it gradually grew in size, being renamed Comacchio Group on 1 November 1983. On that same day the functions of

The Captain General, wearing his Royal Marines tie, about to cut the Corps birthday cake aboard the Royal Yacht Britannia *on 28 October 1977, watched by the detachment and members of the Band. In the centre is Capt (later Lt-Col) Jim Mason. The Royal Yacht Band is part of C-in-C Naval Home Command's Band based in Portsmouth. HMRY* Britannia *also carries a small detachment for duties in the Royal apartments.*

what is now HQ Training, Reserve and Special Forces at Eastney came into being, with its two new roles.

There had been various changes in the Royal Marines Band Service through the years and another came in 1978 when they celebrated their 75th anniversary, with the integration of the Bugler Branch into the Band Service (although they retained their dress distinctions). This year also saw the 25th anniversary of HRH The Prince Philip's appointment as Captain General and he marked the occasion by instituting the 'Prince's Badge', similar to the King's Badge, but awarded to the best all-round musician or bugler passing for duty each year.

Despite the commitment to the NATO flanks, there still remained some scope for action in exotic parts as the final remnants of the old Empire were squeezed dry. The threat of a take over in Guatemala led to the despatch of Royal Marines from ships companies to Belize where they established positions in the dense jungles along the border. Commando company groups had previously deployed to the West Indies on Internal Security duties between 1969 and 1972.

In her Silver Jubilee Year, Queen Elizabeth II reviewed the Royal Marines in Plymouth. On 5 August 1977 she inspected the Guard of Honour commanded by Major Ewan Sale, accompanied by HRH The Prince Philip and the Commandant General, Lt-Gen John Richards, The RM Equerry to the Captain General is Capt Duncan Christie-Miller.

Whilst HRH The Prince Andrew (now the Duke of York) was undergoing Midshipman training at the Britannia Royal Naval College at Dartmouth in the Spring of 1980, he carried out a carefully planned PT syllabus to prepare himself for the commando course. In the second week of March, joining up with the Young Officers batch of September 1979, he completed all the tests required to qualify him for the green beret. This was presented to him on 21 March by the Commandant General, Lt-Gen J C C Richards.

On 27 October 1965 HRH The Duke of Edinburgh, the Captain General, welcomed his uncle, the Earl Mountbatten of Burma, into the Corps when he was installed as the first Life Colonel Commandant at a parade at Eastney. Note that Prince Philip wears the badges of rank of a Field Marshal whilst Lord Mountbatten wears those of a Colonel. Lord Mountbatten carried the unique distinction in March 1942, whilst an acting Vice Admiral, of holding the simultaneous honorary ranks of Lieutenant General and Air Marshal, which he retained till his death.

Jungle warfare has been part of the Corps' armoury for centuries and is still maintained with periodic exercises in Brunei, while 40 Commando spent six months of 1986 as the British presence in Belize.

1979 was a sad year for the Corps when the Life Colonel Commandant, Admiral of the Fleet, the Earl Mountbatten of Burma, who had directed so much of his energy to keeping Royal Marines interests to the fore, was brutally assassinated by the IRA, along with other members of his family, whilst on this fishing boat on holiday in Ireland on 27 August. When he had been installed 14 years earlier he had said:

> I'm proud indeed today that the Royal Marines should feel that I measure up to them and have admitted me to their ranks. Of all the honours that have befallen me, this is the one that I appreciate most and can now share in your 300 years traditions. I look forward to basking in your reflected glory of what the young marines will achieve in the future.

45 Commando provided the Royal Marines Guard of Honour at Westminster Abbey, a marching contingent and street-lining party at his funeral. The Massed Bands preceded the cortège in the procession and another band was positioned along the route to pay tribute to their great champion. It was fitting that a Royal Marines funeral firing party paid the final respects outside Ramsey Abbey. It was perhaps ironic that this brutal assassination should occur whilst 40 Commando were no more than 60 miles away across the border in Londonderry. This was the first time that a Royal Marines Commando had served a full year tour in Northern Ireland, doing so from 1 March

1979. On the whole it was a quiet period for them, but nevertheless a harrowing one for the families left behind.

An unexpected return to Hong Kong in September 1979 was made by 42 Commando who were despatched as an emergency measure when the influx of illegal immigrants from the Chinese mainland into the colony became a flood. Their amphibious prowess was more than appreciated, particularly with the formation of 3rd Raiding Squadron in 1980, equipped with rigid inflatable raiding craft. They operated in the tortuous waterways that separate the innumerable islands. The squadron became a permanent part of the Hong Kong scene and their success rate against illegal immigrants has reduced the latter to a trickle but they are being withdrawn in 1988.

42 Commando returned home after 2 months, hardly expecting to find themselves back in Far Eastern waters within twelve months. On their last day of *Spearhead* duty, insurrection broke out in the New Hebrides, an Anglo-French condominium. A Tactical HQ and 'M' Company Group were flown there on 13 June 1980, restoring the authority of the government without bloodshed or incident. The force, which also included French parachute troops, remained there for two months, taking part in the Vanuatu Independence celebrations on 30 July.

On 17 October 1981 the Corps was stunned when the IRA made an attempt on the life of the Commandant General, Lt-Gen Sir Steuart R Pringle, outside his home in south London. He lost a leg in the attack, but his heroism and defiance have inspired the Corps, making them even more determined during their periodic tours in Ulster. Earlier that year, on 5 May 1981, the strong ties with Norway were further enhanced when Crown Prince Harald of Norway was appointed an Honorary Colonel. He is the first member of a foreign Royal Family to hold such a position in the Royal Marines.

In the continuing fight against illegal immigrants entering Hong Kong, Royal Marines of 3rd Raiding Squadron patrol the tortuous waters by day and night. Here they are seen leaving a Chinese junk after searching it, in one of their rigid raiding craft in 1979.

Lt-Gen Sir Steuart R Pringle gives Cpl Jack Burdett a few tips on walking with an artificial foot. General Pringle was severely injured when his private car was blown up by the IRA outside his home in 1981, and Cpl Burdett was wounded leading his men during 45 Commando's attack on Two Sisters in the Falklands for which he was awarded the Distinguished Conduct Medal. This medal, instituted during the Crimean War was first won by a Royal Marine during the Sudan campaign of 1884. Between then and 1914, 17 medals and two bars were won, 36 more during World War I and a further eight during World War II. Since then only six have been awarded, three in Malaya to Cpl Gillie Howe, Sgt W E Carruthers and TSM Dai Morgan, one at Suez to Cpl D E Mant and one in Northern Ireland to Sgt Gary Pipe.

We have already seen that the Corps maintained a presence in the South Atlantic with NP 8901 in the Falkland Islands and a detachment aboard the ice-patrol ship *Endurance*, which had replaced *Protector*. In 1981 Defence Minister John Nott announced that as an economy measure this ship was to be withdrawn. Argentina, who had long cherished ambitions to take over the islands, saw its chance. Their military government, whose internal policies were bringing economic ruin to their country, seized the opportunity to turn Argentinian eyes away from home. Preparation to invade the Falklands were made and on 2 April 1982 the first Argentinian troops came ashore. This coincided with the arrival in the islands of the annual relief Royal Marines for Naval Party 8901, under Major Mike Norman. It meant that there were 70 marines who rallied to the Governor's call. Hopelessly outnumbered, the garrison put up a spirited resistance and their delaying actions accounted for many of the invaders and some of their

Royal Navy Instructor Officers, known as 'Schoolies', serve with the Corps. During the 1950s and 1960s, when Boy Musicians and Buglers were entered at the ages of 14 and 15, full-time schooling was carried out at the Depot. At one time 15 RN Instructor Officers, led by an Instructor Captain, taught at Deal. Today each Commando still carries one such officer, who, apart from purely educational responsibilities, runs the unit library, is the Resettlement Officer and on operations works as assistant Intelligence and Operations Officer.

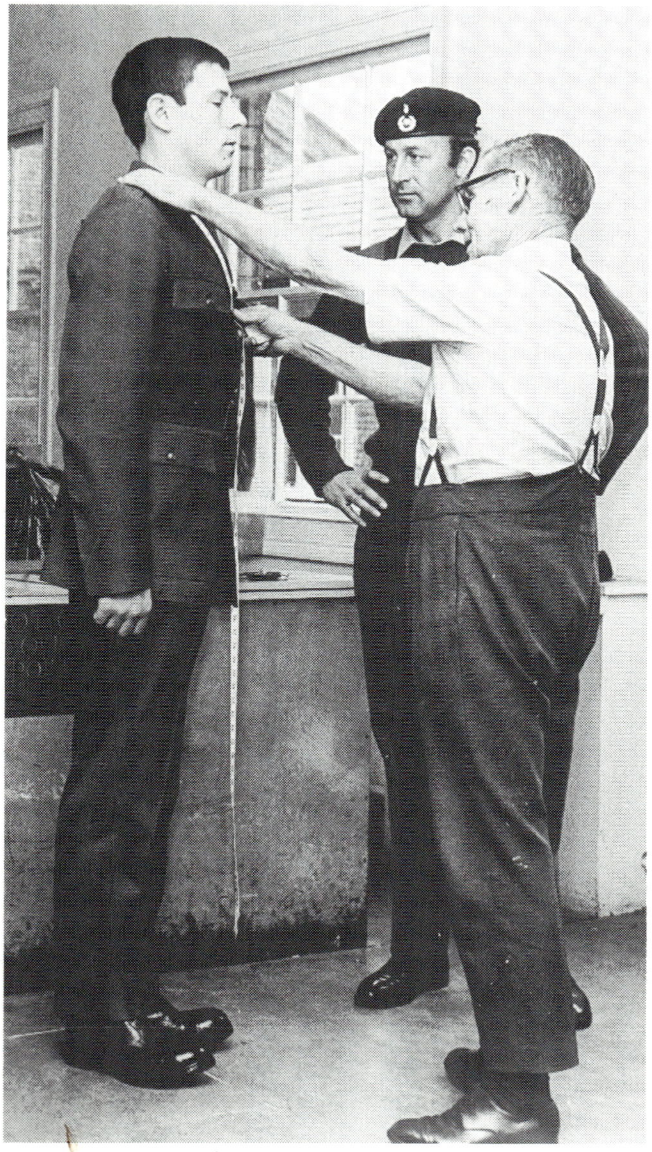

Specially enlisted RM tailors served at all the Divisions and the Depot. They were phased out after World War II when contract tailoring took over. Every recruit was carefully measured and fitted under the supervision of an officer. Lt Chris Menheneott keeps a watchful eye over a new recruit being measured for Lovat uniform at the Depot in 1974.

Capt John Ware conducts the Band of Commando Forces on the helicopter deck of SS Canberra *on its way to the South Atlantic. This band not only provided much appreciated live entertainment for the embarked troops and crew, but during the operation, they acted as helicopter marshals, medical orderlies and guards for prisoners. The Band of Flag Officer 3rd Flotilla carried out a similar task in the SS* Uganda.

armour before the Governor, Rex Hunt, ordered them to surrender.

400 miles to the south-east on 19 March 1982, a party of Argentinians had landed at the old whaling station of Leith on South Georgia and raised their national flag. The Royal Marines detachment of HMS *Endurance* totalling 22 under Lt Keith Mills, were disembarked there on 31 March and set up defensive positions. On 3 April an Argentinian invasion force landed near Grytviken and the Royal Marines fought a two and a half hours battle against overwhelming odds. They even scored a direct hit on an Argentinian corvette with a hand-held anti-tank rocket launcher, before they were also forced to surrender to avoid further casualties. Thus the Falkland Islands fell.

In Britain shock and derision that such a situation could arise rapidly changed to anger, and 3 Commando Brigade was placed at 72 hours notice to sail. Normally seven days mobilisation might be expected, which credits the Soviets with less surprise potential than the Argentinians, but this was achieved as improvised road convoys overcame the inertia of units on two-weeks leave, with marines scattered as widely apart as the USA, Tunisia and Zimbabwe, not to mention Scunthorpe and Inverness. It was to the credit of the Corps that every man answered his unit's recall within 48 hours. The main Naval Task Force sailed from Portsmouth on 5 April with the Headquarters and elements of the Brigade. Many merchant ships and ferries were taken up from trade, the SS *Canberra*, flagship of the P & O Line, embarked 40 Commando (Lt-Col M P J Hunt), 42 Commando (Lt-Col N F Vaux) and 3 Para, who had been placed under command, sailing on 9 April. It was a remarkable achievement, with servicemen and civilians working overtime to prepare and embark a complex of stores, ammunition and equipment needed to fight a war 7,000 miles away. Operation *Corporate*, the recovery of the islands, was under way. While the United Nations waffled, left-wing marchers in London demanded that the Fascist dictatorship in Argentina be allowed to keep the spoils of their aggression.

With shades of the Norwegian campaign of 1940, hastily-assimilated information gleaned from the Plymouth library was pieced together with further

One of the problems of far-reaching amphibious operations is that of keeping the embarked troops fit. During the Falklands campaign three major units, 40 and 42 Commandos and 3 Battalion The Parachute Regiment were embarked in the flagship of the P & O Line, SS Canberra. *Two specially constructed helicopter pads, midships and forward, proved ideal areas for fitness training.*

intelligence as the Task Force steamed southwards. Plans were made on the basis of a remarkably accurate estimate of Argentine reactions by Brig J H A Thompson and his staff aboard *Fearless*. It was thought that the invaders would concentrate the bulk of their forces around the capital of Stanley, and so the sheltered, natural harbour of San Carlos Water, over 60 miles to the west, was selected to establish a bridgehead for the liberating army. It was already known that the enemy would be at least as strong as the British military force, and probably outnumber it. In terms of staff college thinking and planning the expedition should not succeed, but determination, superior planning and training were to be the key factors.

Although extensive physical training took place aboard the ships, it was the short respite of Ascension Island that allowed essential last minute cross-checking of stores, vehicles and personnel to be carried out. The few days each unit were allowed ashore was hardly acclimatisation, as the temperatures in the South Atlantic winter could hardly have been more different, but at least legs could be stretched and weapons test-fired. Meanwhile, as the Task Force steamed further south into heavy seas and icy waters, Argentine air attacks with Exocet missiles proved a blow to morale, though the fleet's protection of the troop-carrying ships was shrewd and consummate. Commando spirits were uplifted when they heard of the recapture of South Georgia by 'M' Coy of 42 Commando under Major Guy Sheridan on 24 April, particularly as the Royal Navy had also damaged one of the few Argentinian submarines in the action.

Considerable intelligence had been collected and disseminated by teams of Royal Marines Special Boats Sections who, with Special Air Service patrols, had already been ashore for more than three weeks. The Task Force arrived off San Carlos on 21 May and the three Commandos, along with the two parachute battalions (2 Para had also come under command by

J and L Companies 42 Commando rest in the deserted and bullet-ridden old sea-plane hangar on the outskirts of Port Stanley, near Moody Brook Camp, at the end of the Falklands War of 1982. The camp had been occupied by the marines of Naval Party 8901 up to the Argentinian invasion on 2 April, when the invaders fiercely attacked what proved to be an empty camp.

this time), went ashore in waves of LCUs to establish a bridgehead at first light. Resistance ashore was negligible but in the attack three Royal Marines helicopters were shot down. The plan was for an advance to be made across the bleak and rugged terrain of East Falkland and take Stanley from the dominant features to the west. This entailed enormous logistic problems and a supply base was set up at Ajax Bay. In their first few days ashore, the ferocity and accuracy of Argentinian air attacks, mounted from the mainland 400 miles away, subjected the Naval Task Force to losses on a scale that forced initial plans to site a main supply base forward at Salvador Water to be abandoned. For the first time in modern history the press, including television, reported actions almost as they took place. This made it very difficult for the families at home, where Wives' Clubs became

Wet and muddy from the fire fight, which at times was so fierce that the whole mountain was illuminated by fire power of various types, Marines of 42 Commando search captured Argentinians after the successful battle for Mount Harriet.

a focal point, and Welfare Services worked overtime to allay the fears of relatives and friends.

The British government were determined that those on the ground must 'get on with it', and political pressure from London was brought to bear on Brig Thompson to break out from the bridgehead. The highly successful, yet costly, attack on Goose Green by 2 Para was a spectacular start. But people in London, who were crying out for more action, could not envisage or even begin to understand the military situation on the barren islands so far away. By this time MGRM Commando Forces, Maj-Gen Jeremy Moore, had arrived to command the Land Forces which now included 5 Infantry Brigade consisting of 2nd Battalion, Scots Guards, 1st Battalion, Welsh Guards and 1/7th Gurkha Rifles. A shortage of helicopter assets forced the brigade commander to order 45 Commando (Lt-Col A F Whitehead), who had sailed south in the *Queen Elizabeth II* and 3 Para, to march eastwards towards Stanley. 45 Commando was in the van to cross the ferocious rolling countryside of tufted scree and ankle-deep bogs, which were interspersed with searing rock-runs. Each man carried the equivalent of at least 110lbs on his back, which included mortar bombs and spare ammunition. The word 'yomp' was coined as the Commando force-marched over 50 miles. The fitness and determination of the marines and paras found them passing through Douglas Settlement and Teal Inlet and reaching their forward positions in half the time predicted by the locals. Very few men fell out. There was a drastic shortage of helicopters to move men forward, although many marines of 42 Commando were in due course air-lifted to Mount Kent for the final assault on the enemy positions defending Stanley. To their chagrin 40 Commando were left in reserve for West Falkland and to defend Ajax Bay.

Forward positions were established on Mount Challenger and Mount Wall. Here the Commandos rested and planned for the Brigade night attack, knowing that the enemy were well dug in, supported by artillery and at least as strong as they were. 45 Commando was tasked to attack the Two Sisters feature in the centre, while 42 Commando were given Mount Harriet on the southern flank. Meanwhile 3 Para prepared to attack Mount Longdon and then Tumbledown Mountain to the north and east. When the two unescorted and unprotected LSLs *Sir Tristram* and *Sir Galahad*, carrying Welsh Guardsmen were bombed with grievous loss of life at Fitzroy, 40 Commando provided two companies to take their place. This delayed the attack until the night of 11 June, and left the Brigade with barely adequate reserves.

In the interim, the opportunity had been taken to scout Argentinian defences and the Royal Marines Mountain and Arctic Warfare Cadre, hastily-formed into a small reconnaissance unit, gave the Brigade commander an added bonus of highly skilled marines. They fought the only major daylight action of the campaign against Argentinian special forces on 31 May, when they caught a group at Top Malo house. Five enemy were killed, seven wounded and the remaining five taken prisoner. But it was the SBS and SAS who silently and unobtrusively gained the vital intelligence which helped Moore and Thompson formulate their plans.

There was little cover on Mount Harriet for any attacking force and Lt-Col Nick Vaux conceived a bold plan of attack involving a covering bombardment from HM ships and field artillery, and a right flanking movement to take the enemy in their left rear. Meanwhile 'J' Company, made up from the original ranks of NP 8901 who had been captured and repatriated, were to cause a diversion on Mount Wall to the enemy's front. Enemy minefields on the line of march added some distance for the attackers. 45 Commando were to assault Two Sisters simultaneously and they chose a left flanking attack from Murrell Bridge. Shortage of ammunition in the ships made the initial bombardment a little less effective

A painting of the famous photograph of 45 Commando yomping into Stanley by Dave Whittley which he presented to Cpl Peter Robinson, whose back view became more famous than his front. Chief Petty Officer Peter Holdgate (left) who took the original photograph recalls that the Union Flag blew off Cpl Robinson's pack into a minefield and was delicately retrieved and remasted before he took the photograph.

Mne Burrell of B Company, 40 Commando rounds up Argentinian prisoners of war at Port Howard on West Falkland. Although 40 Commando were disappointed in not fighting a major battle during the campaign, they did the vital job of guarding the Maintenance Area at Ajax Bay, provided two companies for the Welsh Guards during their night attack on Mount Wall and mopped up prisoners on West Falkland.

than it might have been, but it served to soften up the Argentinians. Any co-ordinated night attack at Brigade level over difficult ground is a risk, but this bold move was made by Brigade Thompson in order to keep casualties to a minimum.

The attack was completely successful, the marines of both Commandos assaulting uphill over treacherous broken country, following taped routes and supported by heavy covering fire. The value of the wire-guided Milan missiles in a bunker-busting role added to the enemy's confusion. 3 Para started the battle for Mount Longdon at 20 00 hours, with the Commandos crossing their start lines at half-hourly intervals. Most leading companies got to within 200 yards of their objectives before being spotted, and then it became a hand-to-hand bullets and bayonets assault. Enemy resistance was stiff from their well dug-in positions amidst the rocky crags of the feature summits, but the careful approach and determined assault quickly overcame resistance. Follow-through attacks on Mount William and Wireless Ridge left the enemy in disarray, and by dawn the British forces dominated the high ground overlooking Port Stanley. The Argentinian Commanders surrendered the port and the islands to Maj-Gen Moore at 21 00 hrs on 14 June. For many of those who had done the main fighting and were still in position on the mountains, the first firm news they had of the surrender came when they heard the BBC World Service news. 'J' Company had the honour of raising the Falkland Islands Union Flag over Government House, the

The Royal Marines Memorial on Plymouth Hoe on which wreaths are laid on Remembrance Sunday each year after a short service. The memorial, which includes two marines representing service ashore and afloat, was unveiled by Earl Fortescue, Lord Lieutenant of Devon, on 8 November 1921 in memory of the 1,449 members of Plymouth Division who lost their lives during World War 1. Designed by W G Storr Barber, it stands at the east end of the Hoe near the Marine Biological Laboratory.

R Commando Company, based at Poole, Dorset, have toured the country visiting schools and giving displays for the past 20 years. Here they gain some national publicity with a death slide from Tower Bridge in London.

Since 1969 Royal Marines have served regularly in Ulster in support of the Civil Power. It is a difficult and dangerous task. Here is a typical scene where Mne Will Manley of 42 Commando makes friends with a young Irish lad during a Crossmaglen patrol.

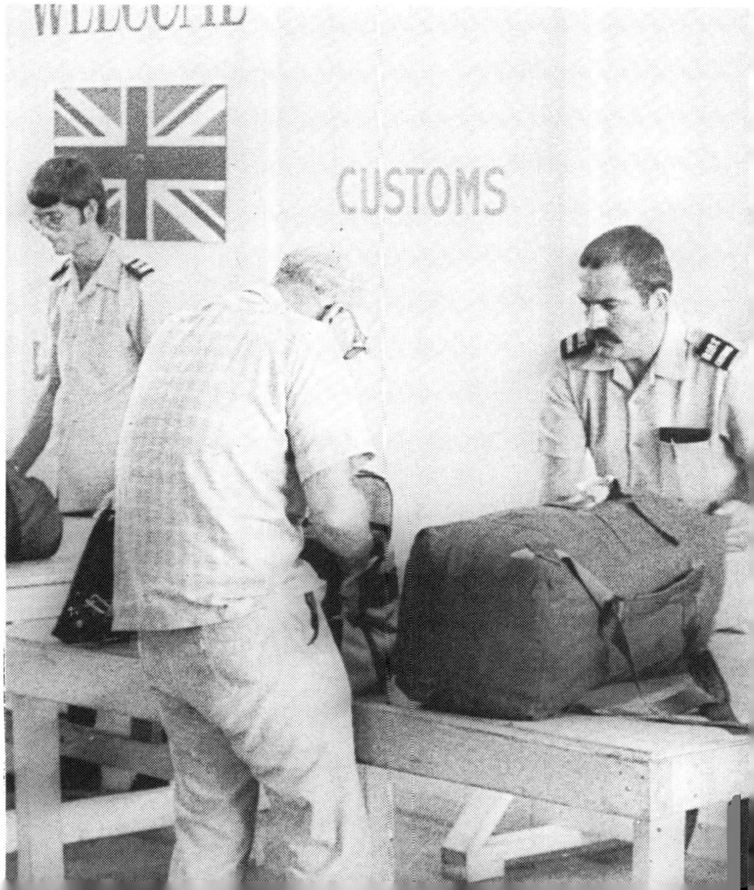

LMA A Jolly and Mne Taff Davies as Customs Officers search baggage at Diego Garcia airport. A Royal Marines detachment has been part of NP 1002 since 1982 when Capt Ken Gill and C-Sgt Terry Sparks took the first Marines to serve in the Maldive Islands in the Indian Ocean since World War II. In addition to providing a British presence, the detachment act as Police and Customs Officers, wearing appropriate uniforms for their jobs.

same flag which had been torn down by the Argentinians only 11 weeks before.

It had been a famous little campaign and a resounding and well-deserved victory. It proved that the skill and expertise of the Royal Marines had not faded: indeed, the efficiency of Britain's Commando force was outstanding. It also highlighted the fighting and campaigning qualities of the Royal Marines as never before, and made those countless gruelling peacetime exercises all worthwhile. The endurance of the Marines in 'yomping' across the island and then fighting a major battle was a tribute to their training across Dartmoor, the Scottish Highlands and in the mountainous terrain of northern Norway. Overall, the British forces had acted with distinction and the only lack was anti-aircraft protection for the fleet, an inherent problem which successive British fleets have seemed unable to overcome. However, the supreme irony of the situation was that had not defence cuts by successive governments so reduced the fleet in the first place, then the war might never have been fought at all. Or, if it was inevitable, it could have been fought rather more efficiently. The ships and warriors returned home to a euphoric welcome reminiscent of the days of the conquest of the Empire.

Whatever the political machinations of the war, the Royal Marines had certainly added another famous laurel to their globe. More than half the Corps had been involved in the South Atlantic and the price they paid was 27 killed and over 100 wounded. However, despite the dizzy welcome home, defence cuts continued in both ships and men.

Throughout this time the Ulster situation had not improved and 40 Commando found themselves back in the troubled province within eight months. By this time HQ 3 Commando Brigade, 42 Commando and the Commando Logistic Regiment had joined 45 Commando in being committed to the northern flank of NATO, while 40 Commando retained their role for out-of-area operations. In October 1983 they sailed for the last time in *Hermes* exercising ashore in Turkey. The last of the traditional aircraft-carriers was laid up in 1984, and though the *Invincible* class carriers have limited accommodation for a Commando Group, this is not their prime function. Despite much discussion renewed entrenchment has led to the abandonment of any meaningful response capability, despite the lessons of the Falklands. The future, however, looks brighter with government promises of more amphibious shipping to supplement civilian ferries, which could be requisitioned in the event of war. Whilst these are adequate for exercises, their military employment is open to considerable question.

Reflective of the inner-looking response today is the Corps' concentration on naval security, even though the Naval Dockyards are privatised. The need to protect off-shore oil-rigs against terrorist attack is also manifest. The RM Reserve today stands at 1,000 officers and men and though Britain may not be able to afford strong defences herself, British expertise was put up for sale during the 1980s and Royal Marines Training Teams were sent abroad to help nations who could afford the best. In the 1950s and 1960s, there were many Royal Marines instructors in Muscat and Oman: subsequently teams served in Iran, Bahrain, Nigeria, Zimbabwe and other countries. This helps many officers and NCOs to keep their hand in at desert, jungle and other forms of specialised warfare, as well as providing interesting and unusual postings abroad. There have been Royal Marines in the United Nations peace-keeping forces in both the Lebanon and Sinai. Throughout 1987 and

Maj-Gen Jeremy Moore, the Commander of the British Land Forces, with the signed surrender document in Port Stanley after the defeat of the Argentine Forces in the Falklands Islands on 15 June 1982. More than half the Corps was involved in this war, either on land, at sea or in the air.

After the recapture of the Falkland Islands in June 1982 and the successful re-occupation of the capital Stanley, Royal Marines from the Commando Logistic Regiment give a new coat of paint to the sign outside the Globe Hotel in Stanley. Perhaps they had their regimental journal The Globe & Laurel *in mind.*

into 1988 Royal Marines have served in the ships of the Armilla patrol in the Persian Gulf where the war between Iran and Iraq has escalated.

In keeping with their 'out-of-area' role, 40 Commando were sent to the British dependency of Belize in late September 1985 for a six-months tour as the resident 'battalion'. The unit was split throughout the country in four different locations, spending most of their time patrolling the border with Guatemala and carrying out jungle patrols to outlying villages. Their contribution to 'Hearts and Minds' was inestimable during this time. Whilst there Lieut HRH The Prince

Sgt Al Harbour served with the Sinai Multinational Force and Observers. He wears the brown beret and badges of the Force. A small number of Royal Marines have been employed in many United Nations Forces such as the Sinai and Lebanon in the 1980s.

Edward, who had joined the Corps as a University entrant in 1983, spent three weeks training with the Commando. He had been the first member of the Royal Family to join the Corps and it was a great disappointment when he decided to resign in 1987.

For the third time in their history, the Royal Marines were invited to carry out London Public Duties in 1986 and the honour fell to 42 Commando to provide guards for Buckingham Palace, St James's Palace, the Tower of London and, for the first time, Windsor Castle.

When Iraqi troops invaded Kuwait in August 1990, Royal Marines were quickly deployed to ships in the Persian Gulf as Air Defence detachments. As the United Nations authorised sanctions, Protection Teams, using rigid inflatable boats to board suspect ships whilst still underway and Lynx helicopters to abseil down, were provided for naval boarding parties. Royal Marines were also deployed to the Allied Headquarters, as security teams for Royal Fleet Auxiliary ships, in Naval Helicopter Squadrons based in Saudi Arabia while the Band of C-in-C Fleet worked as stretcher bearers and medical orderlies aboard the principal casualty reception ship RFA *Argus*. In all 10% of the Corps were deployed.

However the main Corps involvement came after the fighting war had finished in April 1991. The plight of the Kurdish refugees in northern Iraq who had suffered from the retribution being handed out by the Iraqi army became the focus of attention. Operation *Haven* was launched, spearheaded by 3rd Commando Brigade, providing humanitarian assistance and protection to thousands of Kurds who had fled into the mountains. 45 and 40 Commandos, the Commando Logistic Regiment and 1st Amphibious Combat Group of the Royal Netherlands Marine Corps made up the majority of the 4,000 troops from all three Services which were commanded by Major General Robin Ross (MGRM Commando Forces), nominated as Joint Force Commander in Turkey. Royal Marines turned their hands to encouraging Kurdish refugees to come down from the snow covered mountains, and to providing emergency relief with shelter, food, water and medical supplies. By 15 July all forces had withdrawn though one Company of 40 Commando remained as part of the multi-national force in Turkey.

With the collapse of the Communist threat to NATO, the British Services faced a substantial reduction in their size. The Royal Marines took their share resulting in a complete reorganisation at the top. In 1993 the Department of the Commandant General is due to move to Portsmouth and assume the command responsibilities of both MGRM Commando Forces and Training and Reserve Forces, thus concentrating the Royal Marines Headquarters in one location. Meanwhile the Government, appreciating the importance of maintaining an amphibious capability for out of area operations, announced new orders for replacement Assault Ships and LPDs to join the fleet in the late 1990s. This has further strengthened the Royal Marines' links with the Royal Navy.

Thus the Royal Marines continue. Their excellence is unsurpassed, as is their record. Their future in the present political climate of financial stringency is as ever under constant review, but the Corps has survived many crises before, from the days of Charles II through to the 1930s and beyond. Who can doubt that such a proud, skilled and dedicated body of men will survive and emerge even stronger and thus continue to serve the nation loyally for another 324 years? *Per Mare Per Terram* is their motto and their byword. May it long continue.

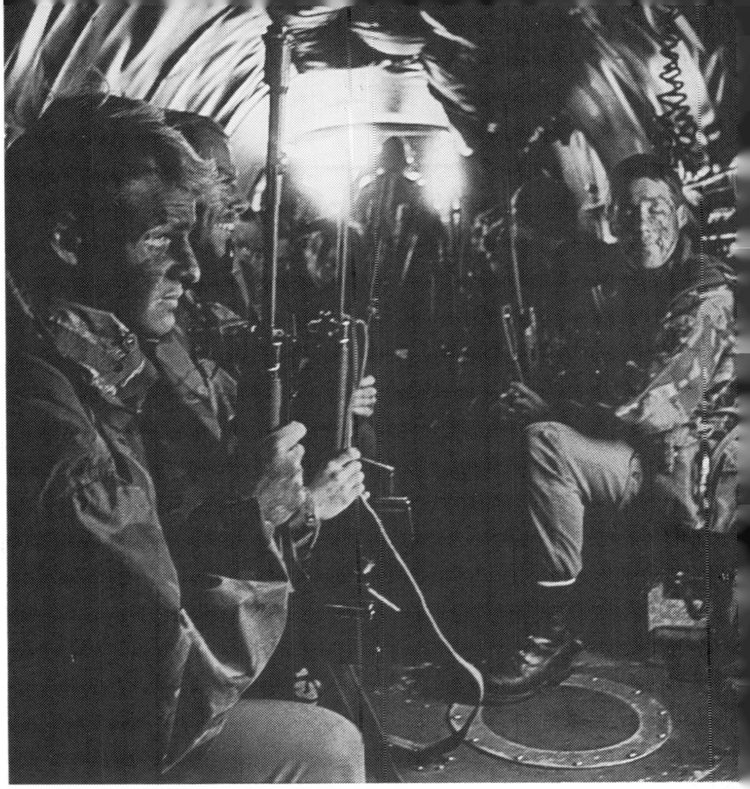

Mne Wayne McGregor of 40 Commando back on patrol in Armagh, Northern Ireland in March 1983 only nine months after losing a foot during the Falklands campaign.

An experience all too familiar with Royal Marines Commandos is the closely huddled expectancy of being landed by helicopter. This 'stick' of 42 Commando are being flown ashore in a Sea King of 846 Naval Air Squadron during exercise Rolling Deep *at Cape Wrath in October 1985. The Sea King helicopter has a crew of four and can carry up to 22 fully laden troops in temperate climes.*

Exercising in Norway Royal Marines Commandos on skis wearing snow oversuits with rifles slung across their chests. Marines are taught to ski and most spend some time on commando exercises within the Arctic Circle.

'Journey Into Space'. Amphibious operations are constantly practised by Commandos when embarked in aircraft carriers. Although the prime functions of CVSs are the anti-submarine and air defence roles, a fully equipped Commando of some 600 men can be embarked. The lift of HMS Invincible *conveys a 'stick' of 42 Commando from the hangar to the flight deck whilst exercising in Northern Scotland in 1985. Note the amount of kit with which the modern marine is required to land.*

Pin-ups and glamour have always played a part of barrack-room life. In 1980 Karen Metcalfe became the first Miss Globe & Laurel, *when she won the title of Butlin's Holiday Princess and photographs of her and her successors have since adorned many a marine's locker. She is seen here aboard HMS* Victory *in Portsmouth with the Royal Marines detachment of 1980.*

A new Commando Memorial was unveiled by Lady Hardy, widow of Gen Sir Campbell Hardy, at Lympstone on 20 September 1986. The imposing statue of two commando figures is by John Doubleday, whose two busts of HRH The Prince Philip and Earl Mountbatten of Burma are on display in the Museum. The memorial commemorates all those who have served in and with Royal Marines Commandos since their formation and incorporates a dedication to General Hardy.

General P X Kelley, the Commandant of the United States Marine Corps, has a 'wet' with Cpl B McDaniel and instructor C-Sgt G Hayes when he visited a Senior Command Course in the field in 1981. There have been exchange officers between the two Corps since soon after World War II, and General Kelley was one of the earliest, qualifying for his green beret in 1960 and subsequently serving with 45 Commando in the Middle East and 42 Commando in the Far East.

RM Buglers sound Last Post at the annual Remembrance service at the Cenotaph in Whitehall in November 1982. Buglers as part of the RM Band Service wear blue capes instead of the khaki great coat, which can be seen worn by D Company of 40 Commando who provided the Guard.

'All Dressed Up For The Lord Mayor's Show', but tradition counts for a lot. Rifles and white Wolseley pattern helmets are propped up against an appropriately signed wall during a break in rehearsals. Royal Marines (and the RM Reserve) have always participated in the Lord Mayor of London's Show, combining this excursion to the capital in November each year with the Festival of Remembrance at the Royal Albert Hall and the ceremony at the Cenotaph in Whitehall. The Wolseley pattern helmet was first introduced into the Corps in 1912.

A popular form of Christmas entertainment, which has raised considerable sums for local charities, is the Sergeants Mess pantomine. None of them would be complete without its 'Corps de Ballet' in the form of the burliest SNCOs complete with tu-tus and hobnail boots.

Royal occasions in London often see Royal Marines lining the streets or providing Guards of Honour. At the wedding of HRH The Prince Andrew to Miss Sarah Ferguson, now the Duke and Duchess of York, in July 1986, L Company of 42 Commando provided the street-lining party and paraded with the SA-80 weapon for the first time on ceremonial duties.

Bandmaster Charles Hotham conducts the Oceanaires, *the official dance band of the RNSM at the RM Association Ball at Seymour Hall in 1948. In his band at this time were Paul Neville (seen standing - later Principal Director of Music RM), Albert Elms (double bass - composer and arranger), John Wagner and Roy Davies (both later Organising Directors of Music RAF).*

The CO, Lt-Col Alan Hooper, with the three Wilkinson Swords of Peace which have been won by 40 Commando, a record for an individual unit in all three Services. The Wilkinson Sword Company has presented one sword annually to each Service since 1967 for fostering good relations in the territories and areas in which they are stationed. 40 Commando won in 1967 for services in Sarawak, in 1972 for Northern Ireland and 1984 for Cyprus. Other RM units to win the Royal Navy's Sword are 41 Commando in 1974 for Cyprus and NP 8901 in 1977 for the Falkland Islands.

Right: Prince's Badge
In 1978, HRH The Prince Philip, Duke of Edinburgh, on his 25th anniversary as Captain General directed that the best all round musician or bugler completing training each year, if of sufficient standard should be awarded the 'Princes's Badge', worn similarly to the 'King's Badge'.

APPENDIX A

CUSTOMS AND TRADITIONS

COLOURS

From the earliest days, Colours were carried by regiments and those of the of the Admiral's Regiment reflected the yellow of their 17th Century coats. It is interesting to note that today 42 Commando still uses the design of the original Lieutenant-Colonel's Colour as their unit flag.

When the six Marine Regiments were formed, the number of Colours carried had been reduced to two, the King's Colour (or Union Flag) and a Regimental Colour (matching the facings). Later, each of the Grand Divisions (Chatham, Portsmouth, Plymouth and, for a period, Woolwich) carried Colours. Each Royal Marine Commando was presented with a stand of Colours at a ceremony in Malta on 29 November 1952, by HRH Duke of Edinburgh, and today in the Royal Marines, Colours are only carried by Commandos. 41 and 43 Commandos, during their brief revivals, were presented with Colours which are now housed in the Officers Messes at Plymouth and Lympstone respectively.

The Queen's Colour is the Union Flag, with the foul anchor and the reigning sovereign's cypher interlaced in the centre. Above is St Edward's Crown surmounted by a scroll with the single battle honour 'Gibraltar'. Underneath is the great globe surrounded by a laurel wreath and below this is a scroll with the Corps motto *Per Mare Per Terram*. The cords and tassels are gold interlaced with the Commando's colour which corresponds with their lanyard. (40-Commando - Light Blue, 42 Commando - White, 45 Commando - Red).

The Regimental Colour is a dark blue flag with a small Union Flag at the pike head, carrying similar central embellishments as the Queen's Colour, with the exception that the cypher of George IV replaces that of the reigning monarch, and the unit numeral is below. The Royal Cypher is at the other corners.

LAUREL WREATH

It is generally accepted that the Laurel Wreath was awarded in recognition of the gallantry of the Marines at Belle Isle in 1761.

THE TITLE 'ROYAL'

Admiral Lord St Vincent, fully recognising the great service which the Corps had given, recommended to King George III that they should be titled 'Royal'. This was granted by Royal Command on 29 April 1802.

THE GLOBE AND GIBRALTAR

In 1827, HRH The Duke of Clarence, who as an Admiral had also been a General of HM Royal Marine Forces since 17 March 1823 and later became King William IV, presented new Colours to all the Grand Divisions. In doing so he announced that HM King George IV had directed that from the difficulty of selecting battle honours amidst so many glorious deeds, the Corps should have the 'Great Globe itself' as its emblem to be surrounded by the Laurel Wreath, retaining 'Gibraltar' as the first great battle honour to represent all previous and subsequent honours.

THE CORPS MOTTO

Per Mare Per Terram - By Sea By Land. It is not known when this motto was adopted, but it first appeared on the caps worn by the Marines at the Battle of Bunker Hill in 1775. It is of interest that the plural of this motto *Per Mare Per Terras* is one of the mottoes of the MacDonalds, and that the son of the Jacobite heroine, Flora MacDonald, served with the Marines in the American War of Indpendence.

THE REGIMENTAL QUICK MARCH

Before 1880 each Division had its own march, which often changed as new Commandants were appointed. The Royal Marines had used '*The British Grenadiers*' and '*The Dashing White Sergeant*' as their regimental quick march and the Royal Marine Artillery adopted '*The Soldier's Chorus from Faust*' until about 1882, when Henry Russell's '*A Life on the Ocean Wave*' was authorised, which incorporated part of a nautical song '*The Sea*'. Henry Epps Sargent, an American, wrote the words, the first verse of which is:

> A life on the ocean wave,
> A home on the rolling deep,
> Where the scattered waters rave
> And the winds their revels keep.
> Like an eagle cag'd I pine
> On this dull unchanging shore;
> O give me the flashing brine,
> The spray and the tempest's roar.

The original arrangement was made by Bandmaster J A Kappey of the Chatham Division. The version used today is by Major F J Ricketts (Kenneth J Alford) and dates from 1944.

THE REGIMENTAL SLOW MARCH

A Regimental Slow March was not officially recognised in the old days, but slow marches were used as inspection pieces. From the time the Corps performed public duties in 1935 *'The Globe and Laurel'*, arranged by Capt (later Lt-Col Sir) Vivian Dunn, based on the traditional old English air *'Early One Morning'* was the Regimental Slow March. It was irreverently known in the Corps as 'The Whore's Lament'! In 1964, the tercentenary year, the Earl Mountbatten of Burma presented the Captain General with Col Dunn's score of *'The Preobrajensky March'* by Donajowsky. This was the march of the Preobrajensky Guards, of which his great uncle the Grand Duke Serge Alexandrovitch of Russia was one of the last colonels. The march had been presented to Lord Mountbatten by King Alfonso XIII of Spain in 1934. After its official adoption as the Corps Regimental Slow March it was first performed when the Massed Bands Beat Retreat on Horse Guard Parade, London, in 1964.

The Captain General receives the score of the Preobrajensky March from the Chief of the Defence Staff, Admiral of the Fleet the Earl Mountbatten of Burma during Tercentenary Year 1964, watched by the Commandant General, General Sir Malcolm Cartwright-Taylor. This march has been used since then as the Regimental Slow March.

THE COMMANDO MARCH

This trekking song of the Boer commandos was often used by Commandos during World War II and was particularly popular with South African officers seconded to the Royal Marines at the time. However Capt Vivian Dunn made an arrangement of it for military band in 1937 after it had been brought to his notice by Major (later Major-General) A N Williams, who had recently served in South Africa. It was officially adopted by the Royal Marines Commandos in 1952. Today it is seldom omitted from the musical programme on ceremonial occasions. The translation from the original Afrikaans is:

Oh take me back to the old Transvaal
That's where I long to be,
I left my little Sarie where the mallies grow
Just by the green-thorn tree.
And I'll be there to meet her
Where I loved her so,
Down by the green-thorn tree.

FREEDOMS

Deal The Freedom of the Borough of Deal was bestowed on the Corps in February 1945, and a Royal Marine was later included as Supporter in the Borough's coat of arms.

Chatham The Freedom was bestowed on the Corps in December 1949, just before the barracks in the city were closed. The last Colours of Chatham Division are laid up in Rochester Cathedral.

Plymouth The Freedom of the city was bestowed on the Corps in May 1955 to mark the 200th anniversary of its association with the people of Plymouth.

Portsmouth The Freedom of the city was bestowed on the Corps in May 1959, when the Captain General accepted it on behalf of the Corps.

Poole The Corps became Honorary Freeman of the Borough and County of the town of Poole in September 1973.

Stanley (Falkland Islands) Because of their unique association with the Falkland Islands, the Freedom was bestowed on the Corps in December 1976.

Exeter Their long-standing connections, going back over 300 years, were recognised when the Freedom of Exeter was bestowed on the Corps in April 1977.

Medway Although the Corps had already received the Freedom of Chatham in 1949, on the creation of the new Borough of Medway (which included Chatham), the Royal Marines were admitted as Honorary Freemen in May 1979.

Exmouth The Corps was adopted by Exmouth Urban District Council in May 1968, and this adoption has since been re-affirmed by the new Borough.

City of London It is often erroneously believed that the Royal Marines have the Freedom of the City of London. No regiment has such a distinction. However, as the Corps has its origins in the Trained Bands of the City of London, and were allowed to recruit from within the city bounds, they enjoy the privilege, along with other selected regiments, of marching through the City 'with drums beating, Colours flying and bayonets fixed'. The Corps was affiliated to the Church of St Lawrence Jewry, Guildhall on 23 April 1974.

ASSOCIATIONS WITH LIVERY COMPANIES

Worshipful Company of Stationers and Paper-makers adopted the Corps in 1949, and the 1952 stand of Colours of 45 Commando are laid up in Stationers Hall, London.

Honourable Company of Master Mariners adopted the Royal Marines Reserve (City of London) in 1953.

Worshipful Company of Armourers and Brasiers make an annual award to the Royal Marine who achieves the best results on the Armourers 3 course.

POSITION IN LINE

A Horse Guards Order of 1820 decreed 'that the Royal Marines, when acting with troops of the Line, shall take their station next to the 49th Regiment.' When the ten regiments of marines, whose numbers were from 44 to 53, were disbanded in 1748, the remaining Line regiments were re-numbered and 49th was the last regiment raised before the Corps of Marines was revived in 1755. However, when the Royal Navy is on parade the Royal Marines take precedence with them on the right of the line.

CUSTOMS

King George VI Candlesticks When HM King George VI dined with the officers of the Corps at the Savoy Hotel on 21 December 1949, each principal officers' mess in the Corps provided a pair of candelabra. On that occasion, the candles were lit just before the loyal toast. HM The King expressed a wish that in future at regimental dinners, after the table has been cleared and the port passed, the lighting should be doused and the King's Candles lit, in what is a ceremony exclusive to the Royal Marines.

Loyal Toast To mark the tercentenary of their formation, HM Queen Elizabeth II dined with the officers of the Royal Marines at the Royal Naval College, Greenwich on 23 July 1964. At the instigation of Admiral of the Fleet, the Earl Mountbatten of Burma (then Chief of the Defence Staff), Her Majesty granted to the Royal Marines when in their own messes the privilege of drinking the loyal toast seated. This privilege extends also to the Sergeants Mess and that of the junior non-commissioned officers.

NICKNAMES AND SAYINGS

Jollies This was the original nickname of the Trained Bands of the City of London, who provided many of the recruits for the Regiment of Sea Soldiers.

Royal Naval mess-deck slang for a Royal Marine.

Lobsters A very old nickname for a soldier and the RMLI (Red Marines from their scarlet tunics) became known as 'Lobsters', while the RMA (Blue Marines from their blue tunics) were 'Unboiled Lobsters', a term seldom heard today.

Bootneck A term deriving its origins from the leather 'stock' which was served out to any soldier to fasten round his neck inside the collar. Bluejackets used to say 'Take my sea boots off your neck', implying that a piece had been cut from his boot to serve this purpose. The expression is now used widely to mean a Royal Marine.

Leatherneck A term derived from Bootneck but normally applied to the United States Marine Corps and the title of their monthly magazine.

Dead Marine Meaning an emptied wine bottle. Lt-Col W P Drury RMLI, a well known writer of naval stories and playwright, relates that the Duke of Clarence used the expression at a dinner party and when a Colonel of Marines looked annoyed, the future King William IV explained that 'He has his duty done once, and is ready to do it again'.

Tell It To The Marines Meaning that only a Marine would be credulous enough to believe anything. There are a number of possible origins. Col Drury told the story that King Charles did not believe his attendants at court when they said that while serving in the South Seas they had seen flying fish. Drury later admitted that this was a figment of his imagination. Nevertheless, the Corps would like to feel that there is enough truth in this concept as Marines who had served all over the world could verify or belie any 'wild' story.

Major A title formerly given in a warship to the Captain of Marines.

Soldier A nickname given by the wardroom to the Subaltern of Marines.

Sgt-Major A title given to the Senior NCO of a ship's detachment. He traditionally reported to the captain of a ship each day that 'all clocks are wound'.

AFFILIATED REGIMENTS

The Queens's Regiment Formed on 31 December 1966 by an amalgamation of a number of Home Counties regiments, including the Queen's Royal Surrey Regiment, which was itself an amalgamation in 1959 of the Queen's Royal Regiment West Surrey, and the East Surrey Regiment. Although both Surrey regiments served as marines during their early history, the Royal Marines particular connection was with the East Surreys who were raised in 1702 as Villier's Marines. The close association was founded on a disaster in 1825 which befell the HQ and half the regiment whilst on passage to India in the East India Company's ship *'Kent'*. Fire broke out during a violent gale in the Bay of Biscay and after abandoning ship the survivors, including many women and children, were transported to Chatham where they were befriended by the Royal Marines. In 1948 the Officers and Warrant Officers Class 1 were authorised to wear the Blue Lanyard, a custom continued in the 1st Battalion of the Queen's Regiment.

The Argyll and Sutherland Highlanders Early connections date from Balaclava in the Crimean War and Lucknow during the Indian Mutiny, but the main association stems from World War II. In July 1940, after the fall of Dunkirk, the 8th Battalion, Argyll and Sutherland Highlanders served with the Royal Marine Brigade for over a year. When HMS *Prince of Wales* and *Repulse* were sunk in December 1941, the Royal Marine survivors joined up with the remnants of the 2nd Battalion, Argyll and Sutherland Highlanders in the defence of Singapore. They formed what became known as 'The Plymouth Argylls', after the association football team since both ships were Plymouth manned. The composite battalion was first commanded by Lt-Col I M Stewart, Argyll and Sutherland Highlanders and subsequently by Capt R G S Lang RM for the last five days before they were taken prisoner. The Corps inter-unit rugby football trophy is the 'Argyll Bowl'.

1st Infantry Battalion (City of Sydney's Own Regiment)(Commando) An alliance with the Royal Marines was approved by HM The Queen in February 1960, whilst No 2 Commando Company of the Regiment has an affiliation with 45 Commando RM.

DRESS DISTINCTIONS

Full Dress Uniform The full dress worn by the present day Royal Marines Band Service is the same as the Full Dress worn by the Corps before World War II.

Band Service Cap Badges Cap badges and helmet plates worn by Royal Marines Bands incorporate features from earlier RM Bands, which were awarded for their service during Royal Tours abroad:

C-in-C Naval Home Command's Band (Portsmouth) has a gilt grenade (dating from the Royal Marine Artillery) on which is surmounted GVR (Tour to India 1911) and a crown, both in silver, all surrounded by a gilt laurel wreath. Above this is the silver cypher EIIR/PP (Commonwealth Tour 1955).

Band of Royal Marines Commandos (Lympstone) wears insignia from the Plymouth Divisional band, the Prince of Wales plume (Canada and Australia Tour 1920) in silver, positioned between the Globe & Laurel and the Lion & Crown.

The Chatham Divisional Band was awarded the silver badge The White Rose of York to commemorate the Duke and Duchess of York's world-wide tour of the colonies in *HMS Ophir* in 1902. The Chatham Divisional band ceased to exist when the Group there closed in 1950.

Corps Colours Corps colours, which form the basis of regimental ties, flags, stable belts, etc, are, in sequence and proportion:

Blue (Navy blue shade) 4 Parts - Marks intimate connection with Royal Navy and blue was the colour of uniform facing for over 100 years.

Yellow (Old Gold shade) 1 Part - Distinctive colour of dress and ensigns of the Duke of York and Albany's Maritime Regiment of Foot.

Green (Light Infantry green shade) 1 Part - Originally worn on shako by infantry of the Corps and perpetuated in bugle strings of the RMLI.

Red (Red drummer shade) 2 Parts - Historic colour of British Army and worn by the infantry of the Corps from the earliest times.

Blue - 4 Parts - As above.

APPENDIX B

CORPS AND UNIT MEMORABLE DATES

The Corps Memorable Dates were revised in 1987 when citations for the Unit Memorable Dates were published for the first time. World War II units, which are still in existence have two memorable dates, one commemorating a wartime achievement and the other a post-war honour.

CORPS MEMORABLE DATES

The Raid on Zeebrugge — 23 April 1918

The 4th Battalion RM under the command of Lt-Col B N Elliot DSO took a leading part in the gallant enterprise against the German Naval Base at Zeebrugge on St George's Day 1918. The RMLI companies landed on the Mole in the face of determined opposition and held their positions whilst the entrance to the canal was successfully blocked and the Mole destroyed. Lt-Col Elliot, the last of a family who had served in the Corps from father to son since 1755, Major Cordner, his second-in-command, 9 other officers and 109 NCOs and men lost their lives in this gallant affair, whilst 233 all ranks were wounded and 13 taken prisoner. Two Victoria Crosses were awarded to the Royal Marines for their conduct during the operation.

Gallipoli — 28 April 1915

During February and March 1915 elements of the 3rd Royal Marine Brigade (Brig C N Trotman RMLI), which was made up by the Chatham and Portsmouth battalions RMLI, landed largely unopposed on the Gallipoli peninsular to dismantle Turkish defensive positions. After the unsuccessful Naval attempts to force the narrows in March the Turkish Army reinforced the peninsular in strength. Thereafter a major amphibious operation was required. The Plymouth battalion RMLI took part in the initial landing on 25 April but the Brigade did not land until the night of 28/29 April when it went ashore at Anzac Cove to relieve 1 and 3 Australian Brigades. On 30 April it was joined in the line by 1 Royal Navy Brigade (Brig D Mercer RMLI) which contained the Deal RMLI battalion. For the next 13 days both brigades were engaged in continuous, heavy fighting bearing the brunt of the Turkish attacks and displaying great resolution. After a counter-attack on the Monash Valley by Chatham and Portsmouth battalions on 3 May 1915, the Turks were driven back with heavy losses. Major Quinn, a great Australian VC, said to Major Jerram of the RM Brigade 'The bravest thing I've seen so far was the charge of your two battalions up that hill on Bloody Sunday.' In another incident L-Cpl R W Parker (Portsmouth Battalion RMLI) was awarded the Victoria Cross for his gallantry in evacuating a party of wounded men under fire.

The Royal Marine Brigade's casualties during this period were 21 officers and 217 men killed, 20 officers and 764 men wounded and 122 men missing. On 12 May both brigades were deployed to Cape Helles to rejoin the Royal Naval Division for the remainder of the campaign.

The Landings in Normandy — 6 June 1944

Over 16,000 Royal Marines took part in the largest amphibious operation in history. Most of the landing craft were manned by Royal Marines and all capital ships carried a RM detachment. Five RM Commandos (41, 45, 46, 47 and 48) were landed during the assault phase; grouped with three Army Commandos into two Special Service Brigades. In addition the Corps provided a number of specialist units including an Armoured Support Group, beach clearance and control parties and Engineers.

The first 48 hours of the operation were the most critical, involving a seaborne assault against a heavily-protected and strongly-held coastline. Most of the RM Commandos were ashore by 0900 on 6 June and had achieved their initial objectives by early on 7 June. The Corps thus played a leading role in the establishment of secure beachheads from which subsequent operations to defeat the Germany Army in the west were developed. Nine officers and 85 men were killed in action on 6 June. The number of wounded is not known. The following gallantry awards were conferred upon Royal Marines during the Normandy campaign, most of them for actions on 6 June: five DSO, three OBE, thirteen DSC, ten MC, twenty-six DSM and thirteen MM.

The Battle of Belle Isle — 7 June 1761

Two battalions of Marines, under the command of Lt-Col John McKenzie, served with great distinction at the siege of Belle Isle near Quiberon Bay. With the 19th Regiment these two units effected the first successful seaborne landing, in the face of stiff opposition. They took part in all the subsequent fighting on the island. The Marine battalions gained great fame at the final storming of the redoubts in June. Of their conduct on this occasion the Annual Register for 1761 said 'No action of greater spirit and gallantry has been performed during the whole war.' The laurel wreath borne on the Colours and appointments of the Royal Marines is believed to have been adopted in honour of the distinguished service of the Corps during this operation.

Recapture of the Falkland Islands 14 June 1982

The Corps was involved in virtually every significant aspect of the South Atlantic campaign, starting on 2 April when Naval Party 8901 opposed the Argentine assault on the islands. A company group from 42 Commando RM recaptured South Georgia on 25 April. From 1 May Special Boats Squadron carried out intelligence gathering patrols which were critical to the success of the main amphibious landing in San Carlos Water on 21 May. The main landing was planned and executed by 3 Commando Brigade RM (Brig J H A Thompson OBE) which deployed at full strength (40, 42, 45 Commandos, RM and Commando Logistic Regiment RM). RM detachments served in many of the ships of the Task Force and all landing craft were RM manned.

On 30 May Maj-Gen J J Moore OBE MC arrived in San Carlos with his Headquarters, based on HQ Commando Forces RM, and assumed command of all land forces including the Army's newly-arrived 5 Infantry Brigades. 3 Commando Brigade RM continued to bear the brunt of the fighting throughout the campaign commanded 5 of the 6 battalion-scale battles that took place before the surrender of the Argentine forces on 14 June. The professionalism and resilience of the Marines who took part were major factors in the success of this unique amphibious operation conducted at a range of nearly 8,000 miles from the UK mounting base. A total of 3,520 Royal Marines, approximately 50% of the Corps, took part in this campaign.

Two officers and 25 men were killed in action during the campaign and 67 were wounded. The following honours and awards were subsequently conferred upon Royal Marines: one KCB, one CB, six OBEs, two DSOs, three MBEs, one DSC, five MCs, two DFCs, ten MMs, one DCM, three DSMs, one DFM and one QGM.

The Battle of Bunker Hill 17 June 1775

On the night of 16 June a rebel American force occupied dominating high ground to the north of the town of Boston where a British garrison was based. On the following morning General Sir William Howe launched an attack to dislodge the Americans, which was repulsed with heavy losses. A second attack was also unsuccessful but the third, after Howe had been reinforced by a Marine Force and the 47th Regiment, finally took the position. The Marine Force under Major John Pitcairn consisted of the First and Second Marine Battalions. It suffered casualties of 29 killed and 87 wounded in storming the heights. A contemporary report said:

> The reputation of the Marines was never more nobly sustained. Their unshaken steadiness was conspicuous and their valour in closing with the enemy when part of the attacking column wavered, gained them not only the admiration of their comrades but the commendation of their distinguished chief.

The Capture of Gibraltar 24 July 1704

The famous attack upon Gibraltar, which led to its surrender to the British on 24 July 1704, was carried out by a brigade of British and Dutch Marines, 1,800 strong under the command of Prince George of Hesse-Darmstadt. In the following October Gibraltar was besieged by the French and Spanish. The Marine brigade, which had been reinforced shortly before by a further 400 Marines from the British fleet, held the fortress against repeated attacks until the siege was raised on 9 March 1705. In one incident in the fighting, Capt Fisher of the Marines with 17 of his men, successfully defended the Round Tower against the continual assaults of 500 French Grenadiers. A contemporary report of this noted defence says: 'Encouraged by the Prince of Hesse, the garrison did more than could be humanly expected, and the English Marines gained an immortal glory.'

The Battle of Trafalgar 21 October 1805

Nearly one third of the Corps was present at Lord Nelson's victory over the combined fleets at Trafalgar, the most decisive sea fight in British history. Ninetytwo officers and over 3,600 NCOs and men of the Royal Marines, at their traditional stations on the upper decks of the British ships, bore a brave and important part in the success of the day. The losses were particularly heavy on board the leading ships; in Lord Nelson's flagship, the *Victory*, all four RM officers and over 50 men of the RM detachment were killed or wounded. The total RM casualties during the battle were 5 officers and 121 men killed or died of wounds and 12 officers and 204 men wounded.

The Birth of the Corps 28 October 1664

King Charles II sanctioned the formation of the Duke of York and Albany's Maritime Regiment of Foot - the first Regiment to be formed specially for service afloat. The Regiment was raised mainly from the Trained Bands of the City of London from which the Royal Marines derive the privilege of marching through the City of London with Colours flying, drums beating and bayonets fixed. The yellow stripe in our present day regimental flash commemorates the yellow uniforms of the Duke of York and Albany's Regiment.

The Assault of Walcheren 1 November 1944

The leading troops in the successful seaborne attack on Walcheren in November 1944, were the 4th

Special Brigade (Brig B W Leicester DSO) consisting of Nos 41, 47 and 48 Commandos and No 4 Army Commando. The three RM Commandos attacked Westkapelle with little support, owing to the weather, other than that provided by the Naval Support Craft, the guns of which were manned by RM crews.

The success of the landing was in no small measure due to the self-sacrifice and gallantry of the Naval Support Craft. After some days' heavy fighting ashore the batteries covering the mouth of the Scheldt were captured. The clearing of the entrance to the river, in which the Royal Marines thus performed a gallant and leading part, was of the greatest importance to the operations of the Allied Armies in Flanders.

This silver centrepiece was commissioned by the officers of the Royal Marines to mark the Silver Jubilee of HM Queen Elizabeth II in 1977. It consists of a modern style Globe and Laurel showing, in nine sections, the countries where Royal Marines had served during the first 25 years of her reign. Operational areas are picked out in gold and the sections represent Western Europe, Eastern Mediterranean, East Africa, Arabian Peninsular, South East Asia (West), South East Asia (East), Korea, West Indies, South America and Antarctica. The other three panels show a Foul Anchor, Commando Fighting Knife and a Lyre representing Sea Service, Commandos and the Band Service.

UNIT MEMORABLE DATES

HEADQUARTERS COMMANDO FORCES ROYAL MARINES

The Recapture of the Falkland Islands — 14 June 1982

HQ Commando Forces was involved in the planning and direction of the South Atlantic campaign from the outset. On the day of the Argentinian invasion the staff of HQ 3 Commando Brigade were in Denmark, 42 Commando had just gone on leave and 45 Commando had half a company in Hong Kong. The task of mounting out the landing force fell to HQ Commando Forces, and the speed with which it was achieved is one of the most remarkable aspects of the campaign. Within 72 hours the Commando Brigade was ready to go to war, reinforced by an Army battalion, a Rapier battery and two troops of armour. Most of the units had been embarked and some were already sailing south.

When it was decided to deploy 5 Infantry Brigade the need for a two star HQ became apparent. No suitable formed HQ was available so HQ Commando Forces had to create one. As its staff were fully committed to providing military advice to the Task Force Commander in his Northwood HQ, no training or work-up was possible before HQ Land Forces Falkland Islands (HQ LFFI) deployed south to arrive in San Carlos on 30 May. HQ LFFI, with no dedicated communications or vehicles, operated out of HMS *Fearless* until the Argentine surrender on 14 June, deploying ashore a small tactical HQ during the last three days of the land campaign. The HQ faced immense problems in providing an *ad-hoc* HQ, and in taking over command in the middle of a complex situation. Its arrival enabled the two Brigades to concentrate on the task of defeating the Argentine land forces, leaving HQ LFFI to deal with Northwood over political and strategic matters, liaison with Naval Commanders, co-ordination of combat and logistic support, overall direction of the land campaign and negotiations for the surrender of the Argentine forces. HQ Commando Forces acting as HQ LFFI thereby made an invaluable contribution to victory in the Falklands.

40 COMMANDO RM

The Landing at Termoli — 3 October 1943

In the early hours of 3 October 1943, 40 Royal Marine Commando (Lt-Col J C Manners) with No 3 Commando and elements of the Special Raiding Squadron landed under cover of darkness at Termoli, a seaport town on the Adriatic Coast, north of the River Bifurno and behind the German lines.

40 Commando penetrated well into the town

before the enemy were alerted. Brisk close quarter fighting with German parachute troops ensued. By 0800hrs 40 Commando had captured the town and controlled the approaches. So complete was the surprise that German vehicles and motor cyclists still drove into a Commando ambush position until noon.

The Germans retaliated in strength and 40 Commando with 3 Commando, the Special Raiding Squadron and some reinforcements from the 78th Division, held off repeated and heavy infantry and armoured counter-attacks by the 26th Panzer Division until eventually the 8th Army linked up with them on 6 October.

The operation was an outstanding success. They had overcome all attempts by a force vastly superior in numbers and armament to dislodge them, and in so doing won a valuable harbour; caused the enemy to withdraw from the natural defence line on the Bifurno and denied them the use of the important lateral road from Naples, thereby forcing them to retreat further northwards.

The Assault on Port Said — 6 November 1956

In the summer of 1956 President Nasser of Egypt nationalised the Suez Canal. The British and French Governments subsequently decided to reoccupy the Canal Zone and 3 Commando Brigade was nominated to spearhead the amphibious assault. At dawn on 6 November 1956, 40 Commando RM (Lt-Col D G Tweed MBE), along with 42 Commando on their right, supported by naval bombardment and close air support, landed across the beaches at Port Said. Its task was to recapture the main part of the town, the Government offices and the docks area. Heavy street fighting followed all day down narrow alleys and through high tenement blocks; all under constant enemy sniping. With determination, military skill and superb junior leadership, all the Commando's objectives were seized before a cease fire was ordered later that day. This was the first major seaborne assault since World War II and 40 Commando played a significant part in this entirely successful phase of the operation.

41 COMMANDO RM

The Landing at Salerno — 9 September 1943

In Italy on 9 September 1943, 41 Royal Marine Commando (Lt-Col B J D Lumsden) landed at Marina in the Gulf of Salerno as part of 2nd Special Service Brigade. The Brigade's task was the capture of the defile of La Molina through which the Allied 5th Army, landing at Salerno, was to pass on its way to Naples 40 miles to the north.

During the climb up a very steep gradient from the beach the Commando surprised German troops, capturing a tank and guns. Strong opposition to the main landing prevented help and relief being sent to the Brigade which was cut off. 41 Commando repelled German counter-attacks for two days, fighting against heavy odds and suffering many casualties. Large numbers of Germans were killed before they finally withdrew and contact was re-established with the main body of the 5th Army

24 hours later heavy German counter-attacks of the bridgehead from high ground overlooking the beaches on another sector caused the Commando to deploy again, and for 5 days it held off continuous enemy attacks which attempted to drive the Allies into the sea.

42 COMMANDO RM

The Battle of Kangaw — 31 January 1945

In Burma during January 1945, following the capture of the Myebon Peninsular, 3 Commando Brigade was given the task of making a further landing near Kangaw, with the intention of cutting Japanese lines of withdrawal down the coast.

On 22 January 1945, 42 Royal Marine Commando (Lt-Col H H Dales) together with No 1 Commando, landed and occupied positions in the mangrove swamp.

Subsequently the Commando was ordered to capture a heavily-wooded ridge known as Hill 170. Two days of hand-to-hand fighting were necessary before the Japanese could be driven from the ridge, and no sooner were they dislodged than they subjected it to heavy artillery fire.

After a lull of several days, the Japanese counter-attacked at dawn on 31 January 1945. The enemy attacked repeatedly. In spite of heavy casualties to the Commando, the Japanese were finally beaten off and withdrew, leaving their dead lying thickly among forward Commando positions.

In a Special Order of the Day to 3 Commando Brigade, Lt-Gen Sir Philip Christison, Commander of XV Corps, concluded, 'The Battle of Kangaw has been the decisive battle of the whole Arakan campaign, and that it was won was very largely due to your magnificent defence of Hill 170.'

The Attack on Mount Harriet — 11/12June1982

During the initial landings on the Falkland Islands on 21 May 1982, 42 Commando RM (Lt-Col N F Vaux) were Brigade Reserve at Port San Carlos, before seizing Mount Kent in a night move by helicopter. By 4 June the unit had moved forward, mostly under cover of darkness, to positions west of the high ground overlooking Stanley and the last Argentine stronghold. After days of probing reconnaissance, a Brigade assault took place on the night of 11/12 June in which the Commando's task was to secure Mount Harriet on the Brigade right flank. By moonlight and in freezing temperatures, 42 Commando moved undetected through enemy minefields in a 9 kilometre right-flanking movement to surprise the enemy in their rear. Consecutive assaults by 'K' and 'L' Companies followed up steep slopes onto company positions amongst the crags at either end of the feature. Careful planning, resolute leadership and the boldness and determination of marines against initially strong resistance and continuous artillery bombardment, eventually prevailed. By first light more than 30 enemy had been killed and over 300 prisoners taken as 42 Commando consolidated on Mount Harriet.

43 COMMANDO RM AND COMACCHIO GROUP RM

The Battle of Comacchio — 2 April 1945

In Italy on 2 April 1945, 43 Royal Marine Commando (Lt-Col I H Riches) was the right flank unit of a 2nd Commando Brigade attack designed to capture the spit of land separating Lake Comacchio from the Adriatic. Against stiff opposition the Commando carried out its task, which involved crossing the River Reno in inflatable dinghies under fire, and established itself by nightfall halfway to the final objective, the canal south of the town of Porto Garibaldi. Next day the advance was continued but the Commando, moving through open country, met intense fire from enemy positions in Porto Garibaldi overlooking the advance. During this fighting 'C' Troop was pinned down in the open by enemy fire. Cpl Tom Hunter immediately engaged the enemy with a Bren light machine-gun and on three occasions from most exposed positions he drew the enemy's fire, thus enabling his Troop to move across open ground to cover. For his bravery, which cost him his life, he was awarded a posthumous Victoria Cross. The Commando secured all its objectives.

In this operation, which had a considerable strategic effect on the campaign, 43 Royal Marine Commando advanced the front line in its area by 7½ miles, and captured 450 prisoners.

44 COMMANDO RM

The Battle of Kangaw — 31 January 1945

In the Spring campaign of 1945 on the Arakan coast in Burma, 44 Royal Marine Commando (Lt-Col G H Stockley) formed part of 3 Commando Brigade. Landings were carried out at Akyab and Myebon to block and cut-off the retreating Japanese and to secure airbases further to the south, in support of the main thrust of the XIVth Army.

On 22 January, a further landing was made some miles inland up the Daingbon Chaung. Here 44 Commando, which had landed in reserve, was ordered forward to secure a hill feature by night attack.

The next evening the Commando, which had meanwhile occupied another hill feature to the east, was counter-attacked strongly by the Japanese. Many casualties were inflicted, but the position was held thus aiding the establishment of the beachhead. On the following day units of 25 Indian Division were landed; they relieved 44 Commando which then withdrew to Hill 170. Subsequently the Japanese began to counter-attack the beachhead, so the unit was withdrawn to the west side of the Chaung to clear the enemy from Daingbon village and to prevent further infiltration from the rear.

In a Special Order of the Day to 3 Commando Brigade, the Commander of XV Corps concluded that the Battle of Kangaw had been the decisive battle of the whole Arakan campaign.

45 COMMANDO RM

The Attack on Montforterbeek — 23 January 1945

During their advance on Linne in Holland on 23 January 1945, 45 Royal Marine Commando (Lt-Col W N Gray) was held up by well-prepared positions behind the Montforterbeek dyke. After hard fighting in bitterly cold weather and over flat ground covered in snow, the leading Troops achieved their objectives. Meanwhile, Commando Headquarters and the remainder of the unit, lying in the snow and frozen with cold, were subjected to heavy shelling from self-propelled guns and suffered numerous casualties.

It was during this action that L-Cpl H E Harden RAMC, the medical orderly attached to 'A' Troop, particularly distinguished himself. He crawled in the snow across 120 yards of flat open ground to dress the wounds of three casualties, and then carried one man back under intense mortar and machine-gun fire before returning with a stretcher party for the other two. Whilst bringing back the third man he was shot through the head and killed. Harden was posthumously awarded the Victoria Cross for his fearless action. The Commander of the 7th Armoured Division passed the following message during the day, which was published as a Special Order:

> To all ranks, 45 Commando. The Divisional Commander congratulates 45 Royal Marine Commando on their valuable work today, which has been of great importance in driving back the enemy from the Divisional front;. Well done Royal Marines! You put up a fine show today and I am very proud of you.

In spite of a determined counter-attack, during which hand-to-hand fighting took place in front of the slit trenches, the captured German positions were held until the unit was relieved by No 6 Commando in the morning.

The Attack on Two Sisters 11/12 June 1982

In the dawn assault on the Falkland Islands on 21 May 1982, 45 Commando (Lt-Col A F Whitehead) landed at Ajax Bay on the Brigade's right flank. After securing the western side of the beachhead, and while ships were being unloaded under Argentine air attack, 45 Commando marched across more than 80 kilometres of rugged terrain in freezing weather and driving rain via Douglas Settlement and Teal Inlet, to be in position on Mount Kent for the main attack by 3 Commando Brigade.

The Commando's objective were the twin peaks of Two Sisters, the centre of the Brigade's three objectives. Bold reconnaissance between 4-9 June by junior leaders had pinpointed enemy positions and fighting patrols, whilst artillery harrassing fire had caused some early casualties to the enemy. A silent approach and a two-pronged attack during the night of 11/12 June against well equipped and dug-in opposition up the jagged, craggy rock formations, culminated in fierce hand-to-hand fighting for the final enemy company positions. About 50 prisoners were captured and 20 enemy either killed or wounded; the remainder having retreated to the east. In the attack, 45 Commando lost 4 men killed and 19 wounded. Thirty six hours later the Commando advanced swiftly to Sapper Hill, again on foot, and thence into Port Stanley.

46 COMMANDO RM
The Attack on Le Hamel and Rots 11 June 1944

In Normandy on 11 June 1944, 46 Royal Marine Commando (Lt-Col C R Hardy) under command of the 3rd Canadian Division and supported by tanks of the Royal Marine Armoured Support Group, were ordered to clear the valley of the River Mue as far south as the villages of Le Hamel and Rots.

The villages were strongly defended from prepared positions by Panzer Grenadiers of the 12th SS (Hitler Jugend) Division. Two troops of the Commando assaulted Le Hamel frontally in line with the utmost determination and in the face of very heavy fire. Two hours of bitter street fighting ensued.

Meanwhile, the rest of the Commando attacked Rots. Although German tanks wiped out the leading section, the Commando fought back courageously against a fanatical foe who contested the advance with the utmost ferocity. Before dusk the remnants of the enemy had been driven from both villages.

This operation, carried out at a time when the Germans were striving to force the Allies back into the sea, greatly contributed to the stabilisation of the Normandy bridgehead.

47 COMMANDO RM
The Capture of Port-en-Bessin 7 June 1944

In Normandy on 'D' Day 1944, 47 Royal Marine Commando (Lt-Col C F Phillips) was tasked with capturing the harbour of Port-en-Bessin, at the western end of the British invasion sector, for use as a supply port.

The Commando, having lost four landing craft sunk by shell fire, landed with the beachhead secured by 50th Division, and then fought their way over ten miles of enemy held country to Hill 72, a prominent landmark two miles South of Port-en-Bessin, where they dug in for the night. Next day, supported by sea and air bombardment, several German strong points were soon captured, but the heavily-garrisoned port held out. At the same time, Germans in a fortified camp south of Hill 72 attacked and overran the Rear HQ of the Commando. The Commando found itself checked in front and threatened from the rear, but as darkness was falling a zig-zag path up to the main strong point east of the harbour was discovered. A successful attack was mounted and the remainder of the garrison surrendered. During the night contact was established with American forces advancing from the west, and in the morning Hill 72 was recaptured by the Commando. Due to courage, determination and self-sacrifice the objective was captured but at the expense of severe casualties.

48 COMMANDO RM
The Landing in Normandy 6 June 1944

In March 1944, 48 Commando (Lt-Col J L Moulton) was formed of officers and men who had recently returned to the United Kingdom after active service in Sicily in infantry and artillery units of the Corps. On 6 June 1944, after only a brief period of intensive training, the Commando landed in assault waves on Normandy. It suffered heavy casualties from close-range fire on the beaches, with two out of its five landing craft sunk.

In spite of the loss of almost half its fighting strength, the Commando reorganised and in the course of the next 24 hours fighting took its objectives, including the formidable strongpoint of Langrune.

COMMANDO LOGISTIC REGIMENT RM
Landing at Ajax Bay 22 May 1982

The Falklands campaign was fought some 8,000 miles from 3 Commando Brigade's base in Plymouth. This stretched the logistic support to its utmost. The Commando Logistic Regiment (Lt-Col I J Hellberg RCT) was faced with the problem of supporting an enlarged Brigade spread over more than 30 warships, auxiliaries and merchant ships. The skill, dedication and exceptional devotion of the ordnance, transport, repair and medical elements of the Regiment in adverse weather conditions, and often under heavy enemy air attack, played a major part in the success of the landings in San Carlos Water. During the following three weeks of the campaign, from their Beach Support Area at Ajax Bay where it had landed on 22 May, the Regiment supported a Divisional Headquarters and two Brigades, treated 695 casualties, processed 2,000 prisoners of war and dealt with over 8,000 tons of stores, ammunition and equipment. The logistic support provided by the Commando Logistic Regiment RM was a battle-winning factor.

3 COMMANDO BRIGADE HEADQUARTERS AND SIGNAL SQUADRON RM

The Landings in San Carlos Water 21 May 1982

After Argentine Forces had occupied the Falkland Islands in April 1982, 3 Commando Brigade RM was the landing element of the amphibious task force ordered to recapture them. The landing force consisting of 40, 42 and 45 Commandos RM, reinforced by 2nd and 3rd Battalions The Parachute Regiment embarked in HMS *Fearless* and *Intrepid*, Landing Ships Logistic and in many ships taken up from trade and sailed 8,000 miles south via Ascension Island. The latter stages of the voyage were conducted in poor weather conditions and under threat of Argentine surface, sub-surface and air attack. Sound planning in conjunction with the Naval Task Force Comanders culminated in a successful unopposed Brigade night landing in the San Carlos region of East Falkland before dawn on 21 May 1982. For the next 7 days the landing force was under constant attack, as the Argentine Air Force attempted to dislodge it from the beachhead. The choice of this remote, sheltered landing area, more than 70 kilometres from the capital of Stanley, enabled the landing force to withstand the air assault and played a major part in ensuring the successful recapture of the Falklands during the ensuing weeks.

3 COMMANDO BRIGADE AIR SQUADRON RM

The Recapture of the Falkland Islands 14 June 1982

From the first landings in San Carlos Water on 21 May 1982 until the Argentine surrender 3 Commando Brigade Air Squadron RM (Major C P Cameron) supported the land forces often under appalling conditions. By day and night in the early stages they were frequently attacked by Argentine fighters and ground attack aircraft. During the initial landings, and in the course of the attack on Darwin and Goose Green, the Squadron lost three aircraft to enemy action, together with four aircrew killed and two wounded. The Squadron was involved in every major ground battle during the campaign in a variety of roles; reconnaissance, liaison, the movement of ammunition to the front line and the recovery of casualties from the forward positions. All these were undertaken in a variety of weather conditions, sometimes at night and often under fire. Its six Scout and nine Gazelle helicopters flew a total of 2,110 hours in just over three weeks, reflecting a remarkable rate of serviceability and flying.

The courage and skill of the aircrew backed by the skill and devotion to duty of the ground support enabled the Squadron to make a significant contribution to the defeat of the Argentine ground forces and to their surrender on 14 June.

OPERATIONAL LANDING CRAFT SQUADRONS

The Landings in Normandy 6 June 1944

On 6 June 1944, Allied forces based in the United Kingdom successfully assaulted the coast of Normandy as a first step to the defeat of the German Army in the west.

Royal Marines manned the assault landing craft carrying the first and subsequent waves of the five leading infantry divisions. In addition, they manned the guns of the support landing craft, and men from the Landing Craft Obstruction Clearance units were among the first ashore to clear the defences on the beaches. For weeks after the initial assault, landing craft crews continued to ferry ashore men, vehicles and stores of the reinforcing divisions.

Both afloat in landing craft and ashore in Naval Beach Parties, the Royal Marines played a prominent and vital part in the invasion.

The Landings in San Carlos Water 21 May 1982

After Argentine Forces had occupied the Falkland Islands on 2 April 1982, a British Task Force was formed and ordered to recapture them. Starting before dawn on 21 May 1982 in San Carlos Water, the Royal Marines landing craft squadrons from HMS *Fearless* and *Intrepid*, together with the 1st Raiding Squadron Royal Marines, landed 3 Commando Brigade onto five separate beaches without loss. After these initial landings the Squadrons continued to offload the logistics in deteriorating weather and under constant air attack. Later all raiding and landing craft were formed into the Task Force Landing Craft Squadron, which subsequently operated on both flanks supporting 3 Commando Brigade and 5 Infantry Brigade in dangerous and testing conditions. The new Squadron also assisted the Commodore Amphibious Warfare in mine-sweeping duties and Special Forces raiding and insertion tasks.

Without this invaluable contribution, before, during and after the main landings, the Falkland Islands Task Force could not have achieved its objective in such a timely manner.

APPENDIX C

ROYAL MARINES VICTORIA CROSSES

These are the official citations as recorded in the London Gazette

CPL JOHN PRETTYJOHN RM
Crimea 2 November 1854

Reported for gallantry at the Battle of Inkerman, having placed himself in an advanced position; and noticed as having shot four Russians.

BOMBADIER THOMAS WILKINSON RMA
Crimea 5 June 1855

Specially recommended for gallant conduct in the advanced Batteries, 7th June 1855, in placing sand-bags to repair the work under galling fire; his name having been sent up on the occasion, as worthy of special notice, by the Commanding Officer of the Right Attack.

LT GEORGE DARE DOWELL RMA
Viborg Baltic Sea 13 July 1855

An explosion having occurred in one of the rocket-boats of the *Arrogant*, during the attack on some forts near Viborg, Lt Dowell (who was on board the *Ruby* gunboat, while his own boat was receiving a supply of rockets) was the first to jump into the quarter-boat of the *Ruby*, and with three volunteers, himself pulling the stroke oar, proceeded instantly, under a heavy fire of grape and musketry, to the assistance of the cutter's crew. The Russians endeavoured to prevent his object of saving the men and the boat, but Lt Dowell succeeded in taking up three of the boat's crew and placing them on board the *Ruby*; and, on his returning to the spot, was mainly instrumental in keeping afloat, and bringing off the sinking cutter.

CAPT LEWIS STRATFORD TOLLEMACHE HALLIDAY RMLI
Peking Legations Boxer Rebellion 24 June 1900

On the 24th June 1900, the enemy, consisting of Boxers and Imperial troops, made a fierce attack on the west wall of the British Legation, setting fire to the West Gate of the South Stable Quarters and taking cover in the building which adjoined the wall.

The fire, which spread to part of the stables, and through which and the smoke a galling fire was kept up by the Imperial troops, was with difficulty extinguished, and as the presence of the enemy in the adjoining buildings was a grave danger to the Legation, a sortie was organised to drive them out. A hole was made in the Legation wall and Capt Halliday, in command of twenty Marines, led the way into the buildings and almost immediately engaged a party of the enemy.

Before he could use his revolver, however, he was shot through the left shoulder, at point blank range, the bullet fracturing the shoulder and carrying away part of the lung. Notwithstanding the extremely severe nature of his wound, Capt Halliday killed three of his assailants, and telling his men 'to carry on and not mind him', walked back unaided to the hospital, refusing escort and aid so as not to diminish the number of men engaged in the sortie.

L-CPL WALTER RICHARD PARKER RMLI
Portsmouth Battalion Gallipoli 1 May 1915

On the night of 30th April/1st May, 1915, a message asking for ammunition, water and medical stores was received from an isolated fire trench at Gaba Tepe.

HM King George V shaking hands with L-Cpl W R Parker VC at Wavetree, Liverpool, during the visit of the Royal Family to attend the consecration of Liverpool Cathedral in 1921. Cpl Parker won his VC as a stretcher bearer at Gallipoli.

A party of non-commissioned officers and men were detailed to carry water and ammunition, and, in response to a call for a volunteer from among the stretcher bearers, Parker at once came forward; he had during the previous three days displayed conspicuous bravery and energy under fire whilst in charge of the battalion stretcher bearers.

Several men had already been killed in a previous attempt to bring assistance to the men holding the fire trench. To reach this trench it was necessary to traverse an area at least four hundred yards wide, which was completely exposed and swept by rifle fire. It was already daylight when the party emerged from shelter and at once one of the men was wounded: Parker organised a stretcher party and then going on alone succeeded in reaching the fire trench, all the water and ammunition carriers being either killed or wounded.

After his arrival he rendered assistance to the wounded in the trench, displaying extreme courage and remaining cool and collected in very trying circumstances. The trench had finally to be evacuated and Parker helped to remove and attend the wounded, although he himself was seriously wounded during this operation.

MAJOR FRANCIS JOHN WILLIAM HARVEY RMLI
HMS Lion Battle of Jutland 31 May 1916

Whilst mortally wounded and almost the only survivor after the explosion of an enemy shell in 'Q' gunhouse, with great presence of mind and devotion to duty he ordered the magazine to be flooded, thereby saving the ship. He died shortly afterwards.

MAJOR FREDERIC WILLAM LUMSDEN DSO, RMA
Western Front April 1917

Six enemy field guns having been captured, it was necessary to leave them in dug-in positions, 300 yards in advance of the position held by our troops. The enemy kept the captured guns under heavy fire.

Major Lumsden undertook the duty of bringing the guns into our lines.

In order to effect this, he personally led four artillery teams and a party of infantry through the hostile barrage. As one of these teams sustained casualties, he left the remaining teams in a covered position, and, through very heavy rifle fire, machine gun and shrapnel fire, led the infantry to the guns. By force of example and inspiring energy he succeeded in sending back two teams with guns, going through the barrage with the teams of the third gun. He then returned to the guns to await further teams, and these he succeeded in attaching to two of the three remaining guns, despite rifle fire, which had become intense at short range, and removed the guns to safety.

By this time the enemy, in considerable strength, had driven through the infantry covering points, and blown up the breach of the remaining gun. Major Lumsden then returned, drove off the enemy, attached the gun to a team and got it away.

CAPT EDWARD BAMFORD DSO RMLI
4th Battalion RMLI Zeebrugge 23 April 1918

This officer landed on the mole from *Vindictive* with numbers 5, 7 and 8 Platoons of the Marine storming force, in the face of great difficulties. When on the mole and under heavy fire, he displayed the greatest initiative in the command of his company, and by his total disregard for danger showed a magnificent example to his men. He first established a strong point on the right of the disembarkation and, when satisfied that that was safe, led an assault on a battery to the left with utmost coolness and bravery.

Capt Bamford was selected by the officers of the RMA and RMLI detachments to receive the Victoria Cross under Rule 13 of the Royal Warrant dated 29th January 1856.

SGT NORMAN AUGUSTUS FINCH RMA
RMA Detachment Zeebrugge 23 April 1918

Sgt Finch was second-in-command of the pompoms and Lewis guns in the foretop of *Vindictive*, under Lt Charles N B Rigby, RMA. At one period the *Vindictive* was being hit every few seconds, chiefly in the upper works, from which splinters caused many casualties. It was difficult to locate the guns which were doing most of the damage, but Lt Rigby, Sgt Finch and Marines in the foretop kept up a continuous fire with pompoms and Lewis guns, changing rapidly from one target to another, and thus keeping the enemy's fire down to some considerable extent.

Unfortunately two heavy shells made direct hits on the foretop which was completely exposed to enemy concentration of fire. All in the top were killed or disabled except Sgt Finch, who was, however, severely wounded; nevertheless he showed consummate bravery, remaining in his battered and exposed position. He once more got a Lewis gun into action, and kept up a continuous fire, harassing the enemy on the mole, until the foretop received another direct hit, the remainder of the armament being put completely out of action. Before the top was destroyed Sgt Finch had done invaluable work, and by his bravery had undoubtedly saved many lives.

CPL THOMAS PECK HUNTER RM
43(RM)Commando Lake Comacchio 3 April 1945

In Italy during the advance by the Commando to its final objective Cpl Hunter of 'C' Troop was in charge of a Bren group of the leading sub-section of the Commando. Having advanced to within 400 yards of the canal, he observed the enemy were holding a group of houses south of the canal. Realising that his Troop behind him were in the open, as the country there was completely devoid of cover, and that the enemy would cause heavy casualties as soon as they opened fire, Cpl Hunter seized the Bren gun and charged alone across 200 yards of open ground. Three Spandaus from the house, and at least six from the north bank of the canal, opened fire and at the same time the enemy mortars started to fire at the Troop.

Cpl Hunter attracted most of the fire, and so determined was his charge and his firing from the hip that the enemy in the houses became demoralised. Showing complete disregard for the intense enemy fire, he ran through the houses, changing magazines as he ran, and alone cleared the houses. Six Germans surrendered to him and the remainder fled across a footbridge on to the north bank of the canal.

The Troop dashing up behind Cpl Hunter now became the target for all the Spandaus on the north of the canal. Again, offering himself as a target, he lay in full view of the enemy on a heap of rubble and fired at the concrete pillboxes on the other side. He again drew most of the fire, but by now the greater part of the Troop had made for the safety of the houses. During this period he shouted encouragement to the remainder, and called only for more Bren magazines with which he could engage the Spandaus. Firing with great accuracy up to the last, Cpl Hunter was finally hit in the head by a burst of Spandau fire and killed instantly.

There can be no doubt that Cpl Hunter offered himself as a target in order to save his Troop, and only the speed of his movement prevented him being hit earlier. The skill and accuracy with which he used his Bren gun is proved by the way he demoralised the enemy, and later did definitely silence many of the Spandaus firing on his Troop as they crossed open ground, so much so that under his covering fire elements of the Troop made their final objective before he was killed.

Throughout the operation his magnificent courage, leadership and cheerfulness had been an inspiration to his comrades.

All ten Victoria Crosses are on display in the Royal Marines Museum at Eastney, Southsea, Hampshire.

The Commandant General, Lt-Gen Sir Peter Hellings, unveils memorial plaques in St Andrew's Church, Eastney in May 1967, to the last surviving Royal Marines holders of the Victoria Cross, Capt (later General Sir Lewis) Halliday (Peking 1900) and Sgt (later Lt) N A Finch (Zeebrugge 1918), who died within a week of each other in March 1966.

BIBLIOGRAPHY

Barker, A J, *Suez - The Seven Day War* (Faber 1964)

Blumberg, Gen Sir H E, *Britain's Sea Soldiers, Vol III* (Swiss, Devonport, 1927)

Blumberg, Gen Sir H E, (with Col C Field), *Random Records of the Royal Marines* (published by the Globe & Laurel 1935)

Bruce Lockhart, Sir Robert, *The Marines Were There* (Putman 1950)

Burns, Ken, (with Mike Critchley), *HMS Bulwark 1948-1984* (Maritime Books 1986)

Cannon, R, *Historical Record of the Marine Corps* (Published 1830)

Crockett, Maj Anthony, *Green Beret, Red Star* (Eyre & Spottiswood 1954)

Edye, Maj L, *The Historical Records of the Royal Marines 1664-1701* (Harrison & Sons, 1893)

Field, Col C, *Britain's Sea Soldiers, Vols I & II* (Lyceum Press, 1924)

Fraser E (with L G Carr-Laughton), *Royal Marine Artillery Vols I & II* (RUSI, London 1930)

Fullick, Roy, (with Geoffrey Powell). *Suez - The Double War* (Hamish Hamilton, 1979)

Globe & Laurel, The, *Journal of the Royal Marines 1893-1988*

Gillespie, Capt Alexander, *Historical Review of the Royal Marine Corps* (M Swinney, Birmingham 1803)

Granville, Wilfred, *A Dictionary of Sailor's Slang* (Deutsch 1962)

Grover, Col G, *Short History of the Royal Marines* (Gale & Polden 1959)

Hampshire, A Cecil, *The Royal Marines Tercentenary* (Paramount Press 1964)

Hampshire, A Cecil, *Just An Old Navy Custom* (William Kimber 1979

HMSO, *The Royal Marines and Their Achievements - 1939-43* (HMSO 1944)

Hollis, Gen Sir Leslie, *One Marine's Tale* (Deutsch 1956)

Jackson, Robert, *Suez 1956 - Operation Musketeer* (Ian Allan, 1980)

Johnson, Maj P K W, *The Story of 46 Commando Royal Marines* (Gale & Polden, 1946)

Jones, Maj A D H, *Short History of the Royal Marines in Deal* (1963)

Ladd, James D, *The Royal Marines 1919-1980* (Janes's, 1980)

Ladd, James D, *Assault From The Sea* (David & Charles, 1976)

Ladd, James D, *Royal Marines of the 1970s and 80s* (Royal Marines Museum, 1981)

Ladd, James D, *Commandos and Rangers of World War II* (MacDonald & Jane's, 1978)

Ladd, James D, *Inside the Commandos* (Arms & Armour Press, 1984)

Marsh, Maj Alan, *Flying Marines* (Royal Marines Museum, 1980)

Moulton, Maj Gen J L, *The Royal Marines* (Leo Cooper, 1972)

Nicolas, Paul H, *Historical Records of the Royal Marine Forces, Vols I & II* (Boone, 1845)

Paget, Julian, *Last Post - Aden 1964-67* (Faber, 1969)

Phillips, Lucas C E, *The Cockleshell Heroes* (Heinemann 1956)

Poyntz, Maj W H, *Per Mare, Per Terram* (Economic Printing & Publishing Coy, 1892)

Pringle, Patrick, *Fighting Marines* (Evans Bros, 1966)

Rose, Col Markham, *Story of the Royal Marines - Lectures for Recruits* (6th Edition published 1932)

Samain, Lt Brian, *Commando Men* (Steven & Sons, 1948)

Smith, Peter C, *Per Mare Per Terram* (Balfour,1974)

Steele, Sir Robert, *Marine Officer Vols I & II* (H Colburn, 1840)

Stokes, E G, *Lower The Ramps - a story of 43 Commando Royal Marines* (Brown, Watson Ltd, 1955)

Thompson, Maj Gen J H A, *No Picnic* (Leo Cooper, 1985)

Trendell, John, *Operation Music Maker - The Story of the Royal Marines Bands* (Published privately 1978)

Warner, Philip, *The Zeebrugge Raid* (William Kimber, 1978)

Young, David, *Four Five* (Leo Cooper, 1972)

Other Unit Histories, Publications and Memoirs

Fuel of the Fire (41 RM Commando) - 1950

History of 41 Commando (Unit History)

History of 43 Commando (Unit History - 1962)

History of 45 Commando (Unit History - 1970)

48 Royal Marines Commando - The Story 1944-46 (Privately published 1946)

40 Commando RM in Haifa - January to June 1948 (Published 1948)

Pay and Records Office Royal Marines (Published 1963)

Memoirs of the late Maj Gen Andrew Burn (Published 1816)

Memoirs of Major Oldfield (Published by G A Pocock, Dartford)

Memoir of Maj Thomas Oldfield by Rev Doctor Mee, (1905)

Memoirs and Services of Lt Gen Sir S B Ellis Edited by Lady Ellis (Published by Saunders, Otley & Co, 1866)

Royal Marines Pocket Books Vols I-IV by Maj P B H Wall and Lt G A M Ritson (Gale & Polden, 1945)

Royal Marines Historical Society Special Publications -

2 - Royal Marines Records 1755-1792

4 - Royal Marines Records 1793-1836

5 - Royal Marines Records 1837-1914

6 - Ghost of a General and the RM Officer Corps of 1914

8 - 41 Independent Commando RM - Korea 1950-1952

9 - Royal Marines in East Devon (History of Lympstone and Dalditch)

10 - Expedition to Siberia - 1919

INDEX

In this selective index, ranks are generally shown as those appearing in the text or caption. Ships are in italics. Army Regiments are shown under their modern titles.